ZEW Economic Studies

Publication Series of the Centre for
European Economic Research (ZEW),
Mannheim, Germany

ZEW Economic Studies

Vol. 1: O. Hohmeyer, K. Rennings (Eds.)
Man-Made Climate Change
Economic Aspects and Policy Options
1999. VIII, 401 pp. ISBN 3-7908-1146-7

Vol. 2: Th. Büttner
Agglomeration, Growth, and Adjustment
A Theoretical and Empirical Study
of Regional Labor Markets in Germany
1999. XI, 206 pp. ISBN 3-7908-1160-2

Vol. 3: P. Capros et al.
Climate Technology Strategies 1
Controlling Greenhouse Gases.
Policy and Technology Options
1999. XVIII, 365 pp. ISBN 3-7908-1229-3

Vol. 4: P. Capros et al.
Climate Technology Strategies 2
The Macro-Economic Cost and Benefit
of Reducing Greenhouse Gas Emissions
in the European Union
1999. XIII, 224 pp. ISBN 3-7908-1230-7

Vol. 5: P. A. Puhani
Evaluating Active Labour Market Policies
Empirical Evidence for Poland During Transition
1999. XVI, 239 pp. ISBN 3-7908-1234-X

Vol. 6: B. Fitzenberger
Wages and Employment Across Skill Groups
An Analysis for West Germany
1999. XII, 251 pp. ISBN 3-7908-1235-8

Vol. 7: K. Rennings et al. (Eds.)
Social Costs and Sustainable Mobility
Strategies and Experiences in Europe
and the United States
1999. VI, 212 pp. ISBN 3-7908-1260-9

H. Legler · G. Licht · A. Spielkamp

Germany's Technological Performance

A Study on Behalf of the German Federal Ministry
of Education and Research

With 31 Figures
and 28 Tables

Physica-Verlag
A Springer-Verlag Company

Zentrum für Europäische
Wirtschaftsforschung GmbH

Centre for European
Economic Research

Series Editor
Prof. Dr. Wolfgang Franz

Authors
Dr. Harald Legler
Niedersächsisches Institut für Wirtschaftsforschung (NIW)
Schiffgraben 33
30175 Hannover
Germany

Dr. Georg Licht
Centre for European Economic Research (ZEW)
L7, 1
68161 Mannheim
Germany

Prof. Dr. Alfred Spielkamp
Fachhochschule Gelsenkirchen
Institut zur Förderung von Innovation und Existenzgründung
Neidenburger Str. 43
45877 Gelsenkirchen
Germany

ISBN 3-7908-1281-1 Physica-Verlag Heidelberg

Cataloging-in-Publication Data applied for
Die Deutsche Bibliothek – CIP-Einheitsaufnahme
Germany's Technological Performance / ZEW, Zentrum für Europäische Wirtschaftsforschung GmbH. Harald
Legler et al. – Heidelberg; New York: Physica-Verl., 2000
  (ZEW economic studies; Vol. 8)
  ISBN 3-7908-1281-1

Physica-Verlag Berlin Heidelberg New York
a member of BertelsmannSpringer Science+Business Media GmbH
© Physica-Verlag Heidelberg 2000
Printed in Germany

Cover design: Erich Dichiser, ZEW, Mannheim

SPIN 10757015   88/2202-5 4 3 2 1 0 – Printed on acid-free paper

Preface

Maintaining the innovation capabilities of firms, employees and institutions is a key component for the generation of sustainable growth, employment, and high income in industrial societies. Gaining insights into the German innovation system and the institutional framework is as important to policy making as is data on the endowment of the German economy with factors fostering innovation and their recent development.

Germany's Federal Ministry of Education and Research has repeatedly commissioned reports on the competitive strength of the German innovation system since the mid-eighties. The considerable attention that the public and the political, administrative and economic actors have paid to these reports in the past few years proves the strong interest in the assessment of and indicators for the dynamics behind innovation activities. The present study closely follows the pattern of those carried out before. It has been extended, however, to include an extensive discussion on indicators for technological performance and an outline of the key features of the German innovation system.

This book results from the joint endeavour of a number of researchers who prepared background reports on various topics which provide data, a thorough interpretation of facts and conclusions. These background papers were presented and discussed in several workshops during the second half of 1998. The researchers who contributed to the reports and discussions on the topics summarized in this book are:

- Marian Beise, Martin Falk, Georg Licht, Friedhelm Pfeiffer, Alfred Spielkamp at the Zentrum für Europäische Wirtschaftsforschung (ZEW), Mannheim,

- Heike Belitz, Alfred Haid, Brigitte Preissl, Dieter Schumacher at the Deutsches Institut für Wirtschaftsforschung (DIW), Berlin,

- Angela Hullmann, Andre Jungmittag, Thomas Reiss, Ulrich Schmoch at the Fraunhofer Institut für System- und Innovationsforschung (ISI), Karlsruhe,

- Birgit Gehrke, Harald Legler, Manfred Steincke at the Niedersächsisches Institut für Wirtschaftsforschung (NIW), Hannover,

- Christoph Grenzmann, SV-Wissenschaftsstatistik im Stifterverband für die deutsche Wissenschaft (SV-WiStat), Essen,

- Steven Casper, Mark Lehrer, Lars-Hendrik Röller, David Soskice at the Wissenschaftszentrum Berlin (WZB), Berlin.

The background papers for this report were prepared by the research team at the NIW on R&D activities of the German industry in international comparison, R&D-intensive industries in Germany, Know-how-intensive services in Germany, international trade in R&D-intensive goods and environmental protection goods, the technological competitiveness of "newly industrialized countries", and the use of human capital in industry. DIW's researchers contributed to the international comparison of non-physical investment, R&D activities in the service sector, industry distribution of R&D activities in industrialized countries, unit values in international trade with R&D-intensive products. The research group at the ZEW prepared papers on innovation activities in German manufacturing and service industries, the role of services in the "national innovation system", and the use and accumulation of human capital and start-ups in high-tech industries. ISI provided background documents on the patent activity in international comparison and emerging industrialized countries, structural trends in patent output, the performance of government-funded R&D facilities and recent trends in selected industries (e.g. medical instruments, pharmaceuticals) as well as in selected high-technology areas (e.g. microsystem technology). SV-WiStat contributed a paper on recent developments of R&D in Germany. The team at WZB prepared case studies on biotechnology and software as well as a paper on the German institutional framework for innovation in international comparision. Marian Beise from ZEW and Heike Belitz from DIW delivered background material on the global R&D networks of companies in Germany. The chapter on innovation policy is based on a background document entitled „areas for action for future innovation policy" jointly written by researchers of DIW, ISI, NIW and ZEW.

Georg Licht, Alfred Spielkamp (ZEW, Mannheim) and Harald Legler (NIW, Hannover) developed the final report based on the background papers. However, the report should be viewed best as the result of a joint endeavour of the whole group of researchers. Therefore, we have to thank all team members for the preparation of excellent background documents, stimulating discussions and insightful comments on various drafts of the report.

Furthermore, we would like to express our gratitude to Bärbel Kahn-Neetix, Johannes Velling, and especially to Engelbert Beyer from the Federal Ministry of Education and Research for their encouragement during the process of writing the report and for stimulating discussions which help to focus the report on innovation policy issues. However, the opinion expressed in this report is the sole responsibility of the authors.

Sarah Zimmer translated the whole report into English. We owe many thanks to her for making the translation process go smoothly. Our special thanks go to Thomas Eckert and Dominik Erlebach (ZEW) as well as to Veronika Machate-Weiß (NIW) for doing a excellent job formating and laying out text, tables and figures. Of course, all remaining shortcomings are our own responsibility.

Mannheim, Hannover, Januar 2000

Harald Legler, Georg Licht, Alfred Spielkamp

Contents

Three forces are driving modern economies – finance, knowledge and social capital. It is no coincidence that all are intangible: they cannot be weighted or touched, they do not travel in railway wagons and cannot be stockpiled in ports. The critical factors of production of this new economy are not oil, raw materials, armies of cheap labour or physical plant and equipment. These traditional assets still matter, but they are a source of competitive advantage only when they are vehicles for ideas and intelligence which give them value.

(Charles Leadbeater, Living on Thin Air, 1999)

Part I   Introduction and Overview

With regard to the international competition carried out between economies and firms, investment in know-how and technological change has become an important determinant of growth and income. Today, the production factor known as "know-how" can be considered as the driving force behind economic development which is accelerated even further by the rise of the service society.

In view of the vital importance of the generation and dissemination of know-how for overall economic growth, income and even employment, the scientific community, managers in the private economy, and policy makers are highly interested in assessments that provide useful information on the innovative power of the German economy. A number of activities go into such assessments: empirical facts must be evaluated on the basis of international comparison, the implications of change in the area of technological performance must be delineated, and the vital foundations for and driving forces behind innovation, know-how-intensive growth and high employment levels have to be ascertained.

Several related assessments of the competitive strength of countries have fuelled the discussion on the perspective of income and employment growth under the conditions of globalisation, the rapid diffusion of new technologies and the structural changes of our economies. Based on the assessment of innovation - related empirical facts are summarized in the 'Green Paper on Innovation' the EU Commission developed their 'White Paper' and their 'First Action Plan for Innovation in Europe' (EU 1995, EU 1997).

More recently, Porter and Stern (1999) have developed a comprehensive indicator on the innovativeness of nations for the leading economies. Primarily based on patent numbers, R&D activity, and GDP per capita, their innovation index points to an increasing advantage for those countries that have a strong medium-term increase in R&D expenditures and/or GDP per capita (e.g. Japan, Finland, Sweden) in the near future.

Contrary to this approach, our study is based on a rich data base comprising various indicators and different aspects and phases of the innovation process. Given the complexity of the innovation process even at the firm level, we are convinced that a global indicator will not adequately reflect the prospects of our economies. Instead, we have to rely on various indicators to guarantee the reliability of the assessment of the present technological performance which can then be used as a guideline to draw conclusions for innovation policy.

Therefore, our approach is more similar to the one used by the OECD in their reports on innovative activities of their member states (OECD 1998a,b,c). However, we concentrate on the German economy, providing and interpreting in detail benchmarking indicators based on more up-to-date data.

In the nineties, Germany's Federal Ministry of Education and Research has commissioned several reports on the competitive strength of the German innovation system. Discussing national economies, these reports use the term "technological performance" rather than "competitiveness" to avoid confusion with competitive relations between firms. These reports have developed a pragmatic definition for this rather vague term and covered a number of vital issues concerning Germany's standing as a location for research and industry with analyses of, *inter alia*, the following questions:

- What is the status of the German industry's research, development and innovation efforts in general, and in light of growing corporate internationalization and the emergence of new competitors from newly industrialized countries in particular?

- How does Germany's research and development performance measure up? To what degree are R&D findings implemented as innovations, new products and processes? What impediments does this involve?

- What standards, strengths and weaknesses do German inventions with world market potential – particularly those involving basic and key technologies – exhibit?

- What strengths and weaknesses can be found in the dynamic forces driving the R&D-intensive sector, its structural change, its integration into international competition and its importance for the level of German industry's performance and macroeconomic results, including findings in respect of Germany's employment situation?

- How important is the service sector as a provider and user of new technologies and a driving force for enlarging industry's store of knowledge and know-how?

- How productive is government-funded science and research?

- How do two major determinants of innovation potential measure up: how high is the level of education among the German population and what efforts do the public and private sectors undertake in the education field?

Technological performance does not change on a regular, year-to-year basis. Instead, the impact that a change in technological performance has on the macroeconomic objectives of greater growth, earnings and employment becomes evident only in the medium and long run. Accordingly, it is necessary to take a more long-term approach when assessing such change. This in turn also implies a continuous approach when observing and assessing the technological performance of a country. The present study is therefore closely linked to its precedents. Although this report on Germany's technological performance does not pay the same attention to some aspects involved, it still offers continuity because the respective analysis always make reference to and draws upon preceding studies.

The report consists of the following three parts:

- **Part II** contains a summary of the study's fundamental findings.

- **Part III** offers observations regarding current challenges to and precepts for education, research and innovation policies. These observations were developed on the basis of the information compiled in the process of conducting the study.

- **Part IV** is devoted to an extensive presentation of the findings and the information obtained in the course of preparing this report on Germany's technological performance in 1998.

Part IV is broken down into several sections: **Section 1** describes the study's "blueprint". The notion that innovation activity is the result of interlinked processes provides the starting point for considerations presented in this section. This section also outlines the various areas that were examined and the corresponding indicators used to measure technological performance. It concludes with an explanation of the basic features of the German innovation system, from which hypotheses for explaining Germany's national innovation pattern are then drawn.

Current data on technological specialization and on comparative advantages in foreign trade with R&D-intensive goods is interpreted in **Section 2**. This is followed by an examination of topical structural issues, using the example of how selected generic technologies have developed.

The service sector's growing importance for technological performance is outlined in **Section 3** using selected indicators. This section focuses on the question of how the interrelationships between industry and services currently look like and what consequences their particular form has for innovation processes in Germany.

Section 4 discusses indicators for technological performance from a short, medium and long-term perspective. This discussion includes developments in patent, R&D, innovation and investment activity and current employment trends in know-how-intensive industries. It also delves into current developments in training requirements and into investment in basic and further training. Attention is directed to the growing danger of a shortage of skilled labor developing in combination with high unemployment among people who have only little or the "wrong" training. This section also examines research activities in universities and non-university research centers to ascertain the levels of commercial exploitation being achieved.

Section 5 is dedicated to special observations. Taking the role of innovation activity in the economic catching-up process as a backdrop, this section begins with an examination of the current level of innovation activity in Germany's new *Länder* and then outlines both actual developments and the development potential of Central and Eastern Europe's countries in transition. This section also pursues the question of how much threat exists for Germany's competitive position in selected fields of technology.

Part II   Summary

The production factor called "know-how" is developing into an increasingly important force for growth and structural change and concomitantly for economic and social prosperity. The ability and willingness of firms to pursue innovation, the conditions underlying the dissemination of new technologies and their translation into new products and production processes, and the existence of a broad knowledge base within an economy play as much a role as do macroeconomic developments and general conditions. The following picture sums up the current status of Germany's technological performance:

The short-term outlook

Looking at the **short-term**, Germany's innovation system displays a relatively high degree of effectiveness: New know-how is quickly translated into patented inventions, innovation activity is increasing viewed (on a cyclically neutral basis), and Germany's ability to assert itself on global markets is substantial. Foreign demand for R&D-intensive goods is a major driving force behind the country's economic growth.

The upward trend in industry's R&D-intensive sector continued through the reporting period. Higher growth rates were reported in cutting-edge technology fields whereas economic sectors in the advanced technology area – Germany's traditional domain – have lost momentum compared to earlier economic cycles (with the automobile industry comprising an exception) and are facing increased price competition.

The medium-term outlook

However, the current upswing does not appear to be strong enough to compensate for the opportunities that industry neglected to use in recent years to expand its know-how base and production capacity. Apparently, **medium-term expectations** on the part of industrial companies are not enough to induce them to significantly expand production capacity in the high-tech sector, regain lost territory in the R&D field or trigger an out-and-out wave of company start-ups. It is this area that gives rise to the first clouds on the medium-term horizon.

Research and development is an investment in technological know-how which can be translated into products and processes in subsequent years. The downward adjustment of R&D activities observed in the industrial sector over the past several years has probably bottomed out in Germany by now. The intensity of corporate involvement in R&D is on the rise once again. This provides a ray of light on the horizon given the fact that R&D activities provide important impetus for technological progress. In R&D-intensive industry, R&D constitutes only a part – which however includes the "essential core" – of all innovation activity.

Nevertheless, when compared on a medium-term basis, the R&D level of German industry is too low, particularly since companies in most countries competing with Germany have returned to investing in R&D with vigor in recent years. Although the medium-term perspective is somewhat subdued, individual fields are seeing a lot of activity: For example, the biotechnology field is doing much to catch up with the international competition, and the automobile industry is busy strengthening its already impressive position. Although the service sector is also sending positive signals in principle, it has nothing outstanding to report on when these signals are measured by international standards. When compared to the USA and other highly developed economies, Germany's service sector still lags behind. Stronger signals from the service sector – particularly the kind that would stimulate innovation in the cutting-edge technology field – would be desirable.

The long-term outlook

In the long run, Germany's schools and universities will decide the question of whether German industry's international competitive strength will last – because education, science and research are key factors of production for an economy's growth, innovation and employment. For this reason, qualified work and a high level of scientific research constitute the best cards that highly developed economies such as Germany have to play in the game of international competition over industrial locations. In the final analysis, innovation is the fruit of investment in education and science. A lack of investment in these areas could eventually hobble innovation, growth and employment.

Looking at Germany, more serious problems are beginning to emerge in connection with these factors – factors which impact technological performance on a long-term basis. The short-term success Germany has had on global markets during the recent upswing and the country's increased innovation efforts should not be allowed to distract attention from this development. Although it can be said that Germany's universities and research institutes are quite productive, they are suffering from a "reform jam," with current performance being the product of earlier investment in these areas. At present, not enough is being invested in the future (education, vocational training, colleges and universities). There would be grounds for great concern were Germany's education efforts to fall even further behind those of its most important rival countries. A projection based on the current international situation would not reveal a very positive picture for Germany.

Deficits in generating new jobs through innovation

Although industry's outlooks have been positive on a short-term basis, R&D-intensive branches have not been able to reassume the role they played during the 1980s – namely, that of a source of new jobs during an economic upswing. The traditional link between growth and employment has weakened, even in R&D-intensive industries. The only new jobs being created – when at all – are for qualified labor. The employment chances of low-skilled workers have deteriorated. As a result, qualification patterns continue to shift.

R&D-intensive industries constitute the core of the German economy's innovation capabilities. They are also the source of efficiency gains which significantly benefit sectors that are located further downstream. However, R&D-intensive industries make only an indirect contribution to solving the country's employment problems. Such industries "bundle" scientific technological know-how and provide solutions which other industries – in the service sector in particular – put to use with the ancillary effect of creating new jobs. At present, **new jobs** are being generated in only a few industries, albeit on a temporary basis. New jobs that promise to be longer-lasting are to be found solely in the service sector where information and communications (I&C) technologies act as driving forces. By international comparison, there is little inter-industry mobility among Germany's work force. Whoever loses his job in the industrial sector today will have a hard time finding a new one in the service sector. The circumstances under which unemployment is to be reduced have not become any simpler.

Weaknesses in Germany's structural shift toward know-how-intensive services

Add to this the fact that although industrial strengths are being maintained, not enough is being done to mobilize new capabilities in the service sector. Despite gains in the service sector, Germany's stock of innovative and know-how-intensive services is still relatively meager when compared with other countries – on a per capita basis, for example. Furthermore, the overall service sector is making little progress with growth or jobs. And it has been unable to offset the job cuts undertaken in the industrial sector in recent years. Germany's service sector lacks sufficient internal momentum. As a result, new, diversified and high-powered service markets are not developing adequately (in contrast to the performance being put in by other countries). Which is also why there is not enough demand being placed on manufacturing to generate innovation.

Unique features of the German innovation system

There is **no** universally applicable, internationally proven "**recipe**" for safeguarding and expanding a country's technological performance. Different countries have different strengths and weaknesses. It is vital that this fact be understood and put to use. Germany's traditional strengths in the advanced technology field would suggest that German firms are successful first and foremost in developing systematic, technology-intensive improvements along well defined trends. This is supported by the German innovation system's ability to use its broad, high-quality know-how base to integrate global know-how from the advanced technology field into the innovation activities of other sectors.

On the other hand, Germany finds it difficult to assume a front-running role in the development of new markets. There is however no reason to consider this to be a fatal weakness. Rather, Germany has simply taken a different route. It is particularly true for Germany that pursuing a "fast-follower strategy" can also bring success. In fact, Germany's broad knowledge base and its regional and sectoral diversity provide the foundation for the rapid diffusion of new technologies and their translation into innovative products and processes and, as a result, for possible catching-up or overtaking processes. For this reason, the crucial question is whether "stragglers" like Germany will always have sufficiently long time frames for translating innovation into value added as the world becomes increasingly globalized.

Education, research and innovation policy approaches

In light of the typical features of the German innovation system and the findings from the report on Germany's technological performance summarized here, Germany's innovation policy should be geared toward:

- Preserving and expanding Germany's position as a global leader in established fields of cutting-edge and advanced technology;

- Supporting the establishment of new competences in new fields of technology so that Germany has a basis for "fast-follower" strategies in various technologies;

- Using education and training policy to step up Germany's transition to becoming a knowledge-based society;

- Increasing the German innovation system's adaptation capabilities and

- Enhancing conditions for innovation activities in Germany.

The German innovation system's strengths are to be found in its regional and sectoral diversity. This diversity can however lead to a mingling of individual minority interests with the tasks falling under the purview of education, research or innovation policy. More attention should be directed to this potential danger. The importance of institutional, direct, indirect-specific and indirect funding has shifted in recent years, and the budgetary leeway available to research and innovation policy has dwindled. Thought should be given to whether these developments should be allowed to continue. Furthermore, establishing clear areas of responsibility would do much to support efforts to increase efficiency within the R&D infrastructure.

Innovation policy measures will have only a modest effect if they are not coherently embedded in Germany's existing innovation system or flanked by complementary reforms. Accordingly, education, research and innovation policies must be integral components of a **broader economic and sociopolitical agenda**.

Education, research and innovation policies must be viewed as cross-sectoral policies

Germany cannot avoid global trends: Responses must be developed to the trend toward ever higher levels of know-how in industry, the structural shifts seen in value-added patterns and employment, and the growing international intermeshment of industry and innovation systems arising from globalization. This entails both opportunities and risks. Germany has the potential to land up in the winner's circle if it makes greater use of its strengths and takes on the challenges facing it. Policies can send signals which can be taken up by science and industry and which can shape public opinion. As a consequence of this circumstance, the political sector has the task of formulating – in conjunction with suppliers and users of technologies, the science community and the economic sector, services and industry – plans and objectives for developing solutions that are marketable on global markets, albeit without prescribing the particular technological means to be used or how these solutions are to be translated into products and processes. This applies above all to the areas energy and the environment, education, health care and the elderly, transport and mobility, as well as to efforts to modernize government. In this sense, "innovation policy" has the **cross-sectoral task** of advocating innovative solutions.

Germany's education, research and innovation policies have an extensive agenda for the coming years. The country's positive short-term prospects should be used to meet the structural challenges awaiting it on the medium and long-term horizon.

Selected Findings

A picture of an economy's technological performance cannot be drawn **using just one "point of departure"**. Rather, a variety of indicators must be used to depict the development of those parts of an economy that are pertinent to an assessment of its technological performance. Only when these indicators are viewed on an overall basis is it possible to evaluate the current situation and prospective trends.

> Both industrial production and productive capacity are geared increasingly toward research, development and the intensification of know-how. Most recently, industry has developed momentum, and production and foreign sales report substantial growth.

Industries in the **cutting-edge technology**[1] field currently **top the growth list** with approximately 7.5 percent of industrial value added. Such industries are posting four percent annual production growth, compared to the three percent reported by industries in advanced technology fields. Production in R&D-intensive industries is expected to grow by eight percent in 1998, approximately twice the level being forecast for other industries.

- **Export demand** has benefited Germany's R&D-intensive industries in particular: Demand from abroad accounted for 90 percent of the **revenue growth** reported by Germany's R&D-intensive industries between 1995 and 1997. Foreign sales by less R&D-intensive industries grew by some five percent between 1995 and 1997. By comparison, foreign sales grew by 17.5 percent in the cutting-edge technology field and by 9.5 percent in the advanced technology field.

- Germany is the world's **largest exporter of advanced technologies** with an 18-percent share of the global market (ahead of Japan at 16.5% and the USA at just under 13%). It also accounts for 11.5 percent of global trade in cutting-edge technologies.

- **Germany's special strength** continues to be in fields that require above-average amounts of high-powered innovation effort ("advanced technology"). Its strength is less pronounced in areas that require extremely high R&D input ("cutting-edge technology"). This is particularly clear when Germany is compared to the USA or Japan because Germany's deficits in cutting edge technology goods are rooted primarily in its trade with the USA and Japan. At European level however, Germany is frequently the technological leader even in cutting-edge technology.

---

[1]   In this study, cutting-edge technology entails goods whose average R&D share is more than 8.5% of turnover. Advanced technology covers goods whose R&D share is more than 3.5% but less than 8.5% of turnover. Together, these two groups comprise industry's R&D-intensive (high-tech) sector.

- The **slight improvement in Germany's position in the cutting-edge technology field** is due first and foremost to the upswing that deregulation has brought to the *telecommunications industry*. By contrast, Germany has seen its foreign trade advantages in *pharmaceutical substances* progressively deteriorate. In the advanced technology field, the *automobile industry* in particular has been posting ever-larger foreign trade surpluses.

- Looking at the **regional pattern** exhibited by German export trade, it turns out that Germany's broad range of products offers a very good match for the enormous demand for capital goods and pollution control goods arising from the modernization and rehabilitation efforts being undertaken by the countries of Central and Eastern Europe.

- The **propensity toward investment** among Germany's high-tech industries increased in 1996 and 1997. Investment in structures and equipment equaled nearly 4.5 percent of turnover in 1996/1997, following a rate of four percent in both of the previous two years. Despite this, production capacity was expanded by only 1.5 percent a year following the recession, a rather meager rate. **Production capacity** had yet to reach 1991 levels in 1997. It has not yet been possible to close the gap in Germany's productive potential that developed during the recession.

- Some 85 percent of the additional industrial capital expenditure that was spent between 1994 and 1999 (or is currently planned) has gone to the R&D-intensive sector. However, **expansion of plant facilities** has just recently gained in importance as a reason for investment, particularly in know-how-intensive industries. This development, together with the expansion of R&D capacity, is a positive signal.

- **Start-up activity** in knowledge-intensive sectors of the economy picked up pace in 1997. The number of new businesses (*Neuerrichtungen*) in **R&D-intensive industries** grew by 1.6 percent in 1997, compared to 0.3 percent in all other industry. As in past years, start-up activity in the **service sector**'s technology-intensive fields (mainly in the area of software production) and business services continued to be particularly brisk, with the number of new businesses growing by approximately 6.5 percent in 1997. Despite this, the share of start-ups in technology-intensive industries out of all start-ups continues to be rather small.

- A disproportionately large number of **university and technical college graduates** are responsible for new start-ups, particularly in the professional services. Currently 19 percent of all university graduates and 13 percent of all technical college graduates are self-employed. People with a degree in science or engineering are likely candidates for founding technology-intensive companies. Given the current trend in the number of students in these fields, skepticism rather than optimism would appear to be called for when developing a prediction for future start-up activity among technology-intensive companies.

- Germany's **self-employment rate** rose markedly during the first half of the 1990s. This growth was due primarily to the increasing number of self-employed persons who have no employees and to a decline in the number of persons in dependent employment. The number of self-employed persons who have employees has been on the decline since 1994. Business services and transport/storage/communications constitute an exception to this trend, with the number of self-employed persons with employees rising in both these areas in recent years. All in all, the number of jobs being created when an individual takes up self-employment is presently lower than in the past.

> Patent activity is on the rise throughout the world. Germany can hold its lead in patent applications with global market potential. Germany's specialization in advanced technologies also extends to its invention activity. There has been little change in the direction of invention activities pursued by major economies.

- Following a lull in the early 1990s, the number of "**triad patents**" for products with global market potential began growing again in 1994. Buoyed by all major industrialized countries, this trend has been due in no small part to the increase in international competition over innovation. A number of smaller economies (such as Switzerland, Sweden and Canada) also reported substantial increases in their patent application levels.

- Measured in terms of population size and economic strength, **Germany generates the most patents** among the world's leading industrialized countries – being approximately on par with Japan and markedly ahead of the USA. However, the USA and Japan have distinct advantages when patent activity is gauged in terms of annual export performance.

- Germany is Europe's undisputed technological leader. Its patent intensity for inventions with global market potential is double that reported by France and Great Britain.

- Germany's patent activity pattern also indicates that its particular advantages are to be found in the advanced technology field. Compared to its level of patent activity in the advanced technology field, Germany's patent activity in cutting-edge technologies tends to be less pronounced. **Specialization patterns in invention activity** have also remained virtually **constant** among the world's leading economies during the 1990s. A few smaller economies (which include Canada and Sweden in particular) have been able to substantially alter the pattern of their innovation activity by making strong gains in various fields of cutting-edge technology.

Industry is expanding its research budgets and spending on innovation activity. However, innovation-related investment continues to lag. Product innovation is gaining in importance whereas process innovation carries increasingly less weight

- **R&D expenditure was markedly increased** across a broad base in 1997, placing it ten percent above the level reported for 1995. As a result, R&D growth outpaced revenue growth for the first time in the 1990s. The **R&D work force** also grew for the first time since 1987, expanding by three percent over its 1995 level.

- R&D alone will not be enough to improve Germany's competitive strength on international markets. Other factors (such as the adaptability of new technologies, and near-market, customer-oriented innovation activity) have become more important for innovation capability and the utilization of growth potential.

- Companies increased the amounts they budgeted for innovation activity for the first time again in 1996 and 1997 (by 9% over 1995), after having cut them back steadily from 1992 to 1995. However, priority has been given to **increasing ongoing** expenditure on the short-term application of know-how to produce new products and processes. By contrast, **investment** expenditure – which tends to be long-term in nature – has picked up pace only **very slowly**; in fact, it has increased at a slower rate than overall investment in fixed assets.

- The amount of corporate turnover generated with **product innovations** (in other words, with products which are new for the respective company) rose markedly during the economic upswing. Today, innovation increasingly targets new, improved products and modern production programs. It is difficult to assess this improvement because shorter product life cycles necessitate an ever-larger portion of turnover to be generated by new products, when one's competitive position is to remain stable. Add to this the fact that competition over product innovation has gotten tougher.

- The increased market-orientedness of innovation activities and the continuation of R&D efforts at stable levels have had a positive effect. The share of turnover attributable to **market innovations** has grown again. However, the number of companies producing market innovations is dwindling. Many product innovations tend to be either gradual refinements and developments of an existing product, product differentiations or imitations. The large share they represent indicates however that technical know-how is being diffused at an accelerated pace.

- Corporate **R&D behavior** and **innovation behavior** are very closely linked on a medium-term basis. In particular, a company frequently uses external know-how to **complement** its own R&D efforts. The ability of small and medium-sized enterprises to work with research institutions and industrial companies on R&D on a collaborative basis increases to the same degree that their involvement in R&D continues at a stable level.

The decline observed in the overall R&D intensity of most industrialized countries in recent years appears to have come to a stop. This also applies to Germany. Government assistance and funding for science and research of increasingly less importance worldwide. There has been little change in the individual countries' respective areas of focus for industrial R&D activities.

- R&D efforts in most major economies remained stable or were reduced in real terms during the 1990s. However, no other country scaled down its R&D as rapidly or as vigorously as Germany did. As a consequence, Germany has **forfeited** the **leading position** it held **in the R&D-intensity rankings** during the early 1990s. Although Germany was able to stop this downward trend and, in some cases, reverse it slightly in recent years, it has not been possible to make up for the deficits that developed during the first half of the decade.

- At 3.6 percent of GDP, Sweden currently leads all other countries in overall **R&D intensity**, followed by Japan, South Korea and Finland (with 2.8% GDP each), Switzerland (2.7%) and the USA (2.6%). Germany reports an R&D intensity of 2.4 percent. It also accounted for 8.5 percent of the aggregate R&D expenditure by OECD countries in 1996.

- **Government involvement** consistently accounts for nearly 40 percent of all R&D spending in Germany and Great Britain. This figure is somewhat higher in France (approximately 47% in 1996), but has begun to fall sharply. The USA has also seen a continual decline in the importance of government R&D involvement since the early 1990s (in the armaments field and large-scale projects). Although government involvement in R&D activities in Japan has traditionally remained at low levels, it was stepped up markedly during the early 1990s and currently accounts for approximately 20 percent of all R&D activity in the country.

- Each country sets different **R&D priorities**: **Germany** concentrates on maintaining a broad base and has traditionally focused its R&D efforts on *advanced technology* fields (automotive, mechanical and electrical engineering, the chemical industry). A shift toward cutting-edge technologies in a number of areas (office machines/EDP, pharmaceutical products, aerospace, measurement and control instruments) has also taken place. In contrast, **France, Great Britain**, the **USA** and even **Japan** funnel large portions of their R&D expenditure into *cutting-edge technology* fields such as the aerospace industry (France, USA), telecommunications (France, Japan), EDP (USA, Japan), and pharma-ceutical products (with Great Britain reporting a marked expansion of its R&D capacity).

- **Small economies** are catching up fast in research and development. The R&D structures found in such countries (Scandinavia, the Netherlands) are generally shaped by a few *large international corporations* which set priorities selectively, primarily in *cutting-edge fields of technology* such as pharmaceuticals, EDP and telecommunications.

> Germany is threatening to drop behind in its investment in education and training. At the same time, the trend toward requiring ever higher levels of qualification from employees continues unabated. The emerging shortage of highly trained personnel who can perform scientific and engineering tasks could turn into an obstacle to economic and technological development in the future.

- Germany's overall expenditure on building and expanding the country's "knowledge base" (in other words, industry's expenditure on the dual education system, continuing education and R&D plus government spending on education, continuing education and R&D) totaled an estimated DM 312.5 billion in 1997. This figure represents 8.6 percent of the country's gross domestic product.

- Public and private sector **spending on education and training** in Germany represents 5.8 percent of the country's gross domestic product, which (following adjustment for demographic factors) is slightly below the OECD average. Although most other OECD countries also reported a renewed decline in this figure, the drop posted by Germany was somewhat larger. It would be grounds for concern if Germany's education efforts were to remain behind those of other highly developed countries competing with it.

- Germany invests an especially large amount in **secondary level education**, primarily due to industry's high level of investment in the country's dual vocational training system. By contrast, the USA and other English-speaking countries invest above-average amounts in **tertiary education**. An international comparison of tertiary level education indicates the existence of structural deficits in the German system insofar as a university degree is very expensive due to the substantial amount of time it generally takes to complete a degree program.

- There are pros and cons to the issue of shorter periods of training. Germany foregoes the advantages of greater occupational and intersectoral mobility arising from shorter **training programs** in favor of providing thorough training that is very specific to the respective occupation. Ongoing **continuing education and training** is becoming increasingly important because the half-life of know-how – the period during which know-how can be economically exploited – is becoming ever shorter.

- Today, some 72 percent of all employees in the manufacturing industry who are liable to social security have completed some form of vocational training.

The share of low-skilled workers (in former West Germany) shrank by nearly five percentage points during the course of the 1990s.

- The demand for highly qualified workers is particularly strong. Today, employees with a college or university degree (and who are liable to social security) constitute nearly eight percent of all employees in German industry, compared to some 5.5 percent in 1990.

- Despite the fact that **unemployment** now affects persons at all qualification levels, investment in education continues to be the best insurance against joblessness: Master tradesmen/technicians and college graduates report five percent unemployment. The rate of unemployment among people who have completed training in an apprenticeable trade is eight percent, undercutting the 18-percent rate registered for people with no vocational qualification by more than half.

- The intensification of services within **industry** goes hand in hand with shifts toward more highly qualified workers. Most notably, the share of **highly qualified employees** in the entire industrial sector has risen from an average of 15 percent in 1990 (former West Germany) to 19 percent in 1997 (Germany as a whole). This growth is due mainly to the increased deployment of scientists and engineers who have training that is essential to technological innovation.

- The **service sector** absorbs an ever-larger number of university graduates, particularly those with degrees in science or engineering. The fast-growing field of business services, banks, insurance companies and transport and communications offer good opportunities for people with degrees in other fields as well.

- International firms have always considered Germany's large, high-quality reservoir of **scientists** and **engineers** to be one of its most important locational advantages. There is the danger that this situation could become inverted should Germany not succeed in making workers with the required skills and qualifications available by some other means. In the short term, it will be impossible to avoid shortages at the start of the next decade because students have already decided what they will be specializing in.

> Germany's scientific and R&D facilities in the engineering and natural science fields generally exhibit a high level of productivity. A positive trend was observed in the area of publications and patents.

- Looking at published research findings in the natural science, engineering and medical science fields throughout the world, Germany generates more than eight percent of the global volume of **publications**, putting it approximately on par with Japan and Great Britain. This figure represents a slight increase in Germany's share of publications – which is understated due to data processing factors. The USA accounts for approximately one third of all publications.

Germany ranks alongside the USA when the yardstick used is the amount of "attention" publications receive, as measured by the number of times findings are cited in international journals.

- Germany has particularly well-developed strengths in **core areas of** physics **and chemistry** (such as solid-state physics/semiconductors, material science, and polymer research) and in measuring and control technology and astrophysics.

- Measured in terms of **patent applications**, universities and non-university research centers have substantially increased the commercial exploitation of their research findings. However, there are considerable differences in their exploitation levels within the R&D infrastructure, even between facilities having comparable functions. Furthermore, not every patent application can be equated with its conversion into a financially successful product. It is a patent's actual exploitation – as reflected, for example, in license revenue – rather than the patent application as such that is material to its actual degree of applicability. When license revenues are used as a gauge, exploitation rates are lower, particularly among Germany's Helmholtz Centers.

R&D-intensive industries are not making much headway in creating new jobs. In tomorrow's knowledge-based society, it is the service sector that will offer employment opportunities. However, even within the service sector, employment is developing differently in various fields.

- The R&D-intensive sector accounted for 2.7 million – approximately 45 percent – of the six million **people working in** Germany's **manufacturing industry** in 1997. This sector generates some 50 percent of value added, indicating that labor productivity in R&D-intensive industries is higher than average.

- A split has developed between **employment** and **production**. Although economic growth and increased employment were still linked with one another and the R&D-intensive sector generated nearly all new industrial jobs during the economic expansion of the 1980s, industry's R&D-intensive sector has not been able to serve as the driving force behind employment in recent years. Only a few areas (the automobile industry and associated fields of electrical engineering; medical engineering; combustion engines/turbines) have been able to produce a positive employment balance in recent years.

- The R&D-intensive sector's direct contributions toward creating new jobs will probably continue to be rather modest through the next several years. The real importance that **R&D-intensive** industries have for economic growth and employment tends to be **indirect** because they represent a substantial portion of Germany's scientific-technological problem-solving expertise. New technologies supply solutions which the service sector puts to use with the ancillary effect of creating new jobs.

- In Germany, **know-how-intensive areas of the service sector** employed some 8.6 million workers (who are liable to social security) in 1997. Which means that just under two-thirds of all jobs in Germany's service sector are now to be found in know-how-intensive fields. These fields account for 35 percent of all jobs in the manufacturing sector (compared to only 30 percent in former West Germany in 1990).

- **Business services** in particular were able to open up new market fields in recent years by using and disseminating I&C technologies. In doing so, they supplied a vital driving force for greater employment in this area. Business services have also profited from the ever-growing demand from industry which has been on the rise since the early 1980s. These trends can be expected to continue.

- By contrast, employment is currently on the decline in the **banking and insurance industries** due primarily to rationalization efforts which center on increasing the use of I&C technologies. There is every indication that this trend will pick up more steam as competition grows.

> Industry and the service sector stimulate each other's innovation activities. Innovation is becoming increasingly important in the service area as well. However, government regulation hampers the utilization of innovation capabilities.

- **Performance** in the **industrial sector** is dependent upon the **service sector** and vice versa. German industry's strong position on international markets would be virtually incon-ceivable without the existence of high-powered services at its doorstep. New "clusters" are emerging in growing numbers between innovative industrial companies and service providers. Examples of this would include the close links between medical technology/the pharmaceutical industry and the health care sector, between telecommunications and telephone corporations, and between aeronautical engineering and airlines.

- The level of overall **momentum** in the **service sector** has however left much to be desired in recent years, particularly in regard to developing new markets. Although the service sector has taken on more weight in relative terms over the past several years, Germany's stock of innovative and know-how-intensive services is still relatively meager when compared with other countries – on a per capita basis, for example. As a result, industry is not being pushed to be innovative. The service sector's "lead market" function is frequently not being put to enough use in Germany.

- Innovation potential often cannot be put to use because **services** are more highly regulated in Germany than in many other countries, with the general effect of putting a damper on innovation activity. Foreign investors in particular bemoan Germany's thick regulatory maze and the requirements it imposes on market access in precisely those parts of the German service sector that tend to be geared to "public needs" (such as health care, the power supply, transport and public utilities).

- There is an unmistakable trend toward **standardizing** services (in other words: toward the "industrialization of services"). This trend is leading to a further increase in the international division of labor in services. Innovation activity is becoming increasingly important for export performance in the service sector as well.

- As a force behind innovation, **R&D activities** play a much smaller role in the service sector than in the industrial sector. They are concentrated in the software, transport and telecommunications fields. Research and development are generally less technology-oriented and less formalized in the service sector than in other sectors. Technological developments in the industrial sector often provide the impetus for developing or modifying services.

> Germany's new *Länder* are slowly making progress with their integration into international markets, but are – all in all – still lagging considerably behind the progress reported by former West Germany. Progress is being made in stabilizing the region's R&D capabilities.

- The increased activity observed in east German export trade is a positive sign. Still, **east German firms** generated only 25 percent of their turnover with R&D-intensive exports in 1997, compared to the approximately 50 percent reported for all firms in Germany. Based on this, east German companies accounted for only 2.5 percent of Germany's total **foreign sales** by R&D-intensive industries in 1997. The R&D-intensive sector in Germany's new *Länder* accounts for approximately one third of the region's **industrial production**, making it considerably less important than it is in the western *Länder*.

- The number of **firms conducting R&D** has risen in recent years and **R&D spending** is likely to have increased slightly in 1997. The number of **patent applications** being generated by East German companies is growing considerably faster than the rate reported for Germany as a whole, albeit from a baseline that is approximately one third of the national total. The number of businesses being newly established in the new *Länder* has diminished following several years of rapid growth which lasted up until the mid-1990s. This is also the case in R&D-intensive industrial fields. The new *Länder* account for a smaller share of the country's total business services; however, using this baseline, the rate at which new businesses are being established in this area is approximately as high in the new *Länder* as it is in former West Germany.

A country's national innovation system provides a picture of that particular economy's specific strengths and weaknesses in the international contest over innovation. In the German innovation system, companies are especially successful in developing cumulative, technology-intensive improvements that follow well defined trends. Germany is initially hesitant in its response to new technological challenges but then becomes all the more actively involved in those segments that match its innovation patterns well.

- Factors that play a decisive role in the technological and economic success of innovation include not only education and research efforts, but also the infrastructure, the credit and capital markets, government regulation, state demand for innovative goods and services and, last but not least, market potential. Seen from the angle of an international comparison, **Germany's innovation pattern** – high-quality innovation that follows clear trends – corresponds to a number of important institutional conditions: availability of long-term capital, cooperative trade unions, influential employers' associations, smooth-running training systems, strong employee loyalty and close collaboration between companies.

- Germany has also specialized in those areas of the cutting-edge technologies **biotechnology** and **software** that offer good matches for its strengths: *platform technologies* and *software services*. Cumulative product and process development that follows the "lead user's" example is especially promising in these areas. German firms have acquired broadly applicable competence in such areas. By contrast, their presence is less pronounced in the therapeutical agent branch of the biotechnology field and in the "off-the-shelf" standard software field.

- **Microsystem technology** is considered to be a potential key technology for the coming century. Germany has currently acquired a *good starting position* for itself in microsystem technology, both as a science and as a technology. In coming years, it will be essential that more be done than in the past to link microsystem technology with those fields in which German firms have traditionally exhibited strength.

- In the area of **environmental engineering**, Germany can make particularly good use of its typical strengths in translating existing know-how and in positioning itself on the market. Germany shares honors with the USA as the world's leading exporter of environmental technology. However, weak demand and a lack of challenges for environmental innovation have clearly left their mark on innovation activity among German firms over the last several years. New incentives to invest in environmental protection and the resultant market potential this would create in Germany would also enhance the international competitive strength of manufacturers of (additive and integrated) pollution control products.

- The risks entailed in strategies that are geared to catching-up processes, overtaking processes and widespread impact have grown as product and technology life cycles have become shorter and structural change proceeds at an ever more rapid pace. The players in the German innovation system must therefore exhibit greater willingness to take risk in the future so that they are able to respond more flexibly to technological, occupational and sectoral structural change.

Part III Challenges to and Precepts for Effective Education, Research and Innovation Policies

Germany's future will be built on education, science, research and technology. These fields comprise Germany's "traditional" strengths and a critical foundation for providing for the future. German companies are often a step ahead in the race to win the customer's favor. However, the frontrunners' edge is dwindling, know-how leads have increasingly shorter half-lives, and the pressure to be more productive is growing, concomitantly increasing the necessity of making know-how marketable within ever-shorter cycles.

There is **no** universally applicable, internationally proven "**recipe**" for safeguarding and expanding a country's technological performance. Different countries have different strengths and weaknesses. It is vital that this fact be understood and put to use. Germany's national innovation system is often described as a "model of cooperative consensus" in which innovation and new technologies are developed and become established in a kind of coordinated team effort between science, research, business, government, trade unions, banks and federations. Above all, it is important that attention be paid to the interdependencies within this system: Weaknesses in one area will restrict other areas and limit avenues for action.

The German innovation system in the global arena

By international standards, Germany's education, research and innovation policies exhibit elements that are more strongly **diffusion-oriented** than those of other countries. These policies strive to establish the broadest possible technical, sectoral and regional base for industry's innovative capacity. Germany's traditional **strengths** in the advanced technology field would suggest that German firms are successful first and foremost in developing systematic, technology-intensive improvements along well defined trends.

- The German innovation system's ability to use its broad, high-quality know-how base to integrate global know-how in the advanced technology field into the innovation activities of other sectors supports this approach. Essential factors to the success of these efforts are the broad diffusion of new technologies in conjunction with a basic and continuing education and training system that is geared to educating broad sections of the population.

- On the other hand, the German innovation system finds it difficult to assume a front-running role in the development of market fields that are based on new technologies. German industry often takes a certain amount of time before it absorbs innovations generated in the cutting-edge technology field.

There is however no reason to consider this to be a fatal weakness in the German innovation system. Rather, this scenario describes **a different path**. The latest developments in the biotechnology field show that when the capabilities of the German innovation system are adeptly combined with the requirements of new technologies, it is actually possible to jump onto a "moving train." It is particularly true for Germany that pursuing a **"fast-follower strategy" in new market fields** can also bring success. However, education, science, research and technology do not diminish in importance when such a course is taken. In fact, Germany's broad knowledge base and the regional and sectoral diversity of its innovation system provide the foundation for the rapid diffusion of new technologies and their translation into innovative products and processes and, as a result, for possible catching-up or overtaking processes.

At the same time however, Germany must preserve and expand its **position** as a **leading** innovator in **established fields of technology**. Besides being necessary for safeguarding growth and employment, this is also a key prerequisite for rapidly integrating new innovation fields into Germany's "portfolio." Particularly when these fields are increasingly compatible with the traits of Germany's innovation system, opportunities arise – for a short time – for German companies to tap them. The full strength of Germany's innovation system can be brought to bear when it is possible to make use of established industrial networks between manufacturers and suppliers.

Reasons for modern education, research and innovation policies:

- Imperfect markets and externalities impair the processes involved in generating, applying and diffusing new know-how. Action on the part of government can help mitigate the effects of market failure and market imperfections.

- As the result of lock-in effects, companies are too late in responding to new technological developments and in initiating new technological trends. Government measures could possibly help overcome lock-in effects.

- Modern policies could help avert the danger of insufficient investment in the R&D infrastructure and promote investment in the I&C infrastructure (with the goal of stimulating the development of innovative services in particular).

- Rapid technological development, such as in the field of I&C technologies, leads to a swift decline in absorbing capacity and, as a result, reduces the scope for adaptation (particularly for small and medium-sized enterprises); training policy and support for lifelong learning processes help companies keep pace with technological progress.

Government innovation policy should also pay attention to ensuring that the market's efficacy as a selection mechanism is sufficiently fostered. A selection mechanism that is too weak is just as dangerous as one that is overly strong and does not give new companies with new product ideas a chance, and thereby reduces the diversity of an economy's options for innovation on a medium-term basis.

- And finally, one of the key functions of government innovation policy is to ensure that the complementarities within the German innovation system bear fruit. An effective and efficient government innovation policy demands the continual evaluation and, when necessary, adjustment of the government-funded R&D infrastructure. Only in this way is it possible to respond to new technological developments in a timely manner and ensure – by modifying existing rules and institutions within the German innovation system – that corporate innovation efforts are adequately fostered.

Several elements of the German innovation system prove to be a **disadvantage**, particularly in regard to being able make a rapid transition to new markets: Although the corporate financing system's long-term stance, the high level of investment in occupation-specific or industry-specific know-how and the traditionally extensive coordination and collaboration between the players in the innovation system facilitate optimization within the system, they also restrict flexibility. For this reason, the German innovation system is sensitive to technological shocks and sudden radical change. For example, shorter product and technology life cycles mean less leeway for adjustment. They also up the risk of not being able to carve out a niche in new market fields in time. Players in the German innovation system must therefore exhibit **greater willingness to take risk** in the future. This in turn requires the prospect of higher returns on investment – in human resources as well as in real capital – in the future.

Long-term interaction between **all** parties involved in developing an innovation is characteristic of successful innovation processes. Feedback and learning effects are essential elements in any innovation. It is these factors that make an innovation system dynamic. As a result of internationalization, the strategic orientation of multinational companies, financial interrelations and the cross-border transfer of know-how both within Europe and beyond its borders, the German innovation system is not a closed system which global trends bounce off of. These new challenges are putting the system's flexibility to the test. All players in the system are responsible for establishing the conditions necessary for ensuring that Germany will be able to profit in the future as well from its transition to being a knowledge-based society.

(A) Use education policy to strengthen innovation capabilities.

The question of whether German industry and German jobs will be able to hold their own in the international arena in the future will be decided by Germany's schools and universities.

Human capital is **key** to maintaining a know-how-intensive society's competitive strength in the medium and long term. Consequently, Germany cannot allow investment in education to flag. As a matter of fact, those government budgets that fail to recognize the economic nature of investment in education and training and therefore treat it as a consumer good must assign it higher priority. This applies to all local and regional authorities. Although cuts in education spending might be considered "savings" today, they actually reduce the amount spent on provisions for the future. This will, in turn, lead to lower income levels and tax revenues in coming years. The right choice would be a **two-track strategy** that couples reform with greater efficiency and funding hikes to increase the education system's productivity (see Section 4.3.1 and Section 4.3.2).

An individual with a sound general education will have greater job security in the future and a better foundation for "lifelong learning."

The period during which training can be economically exploited – training's **half-life** – is dwindling. Specialization and, concomitantly, continuing education and training needs and/or specialized qualification continue to increase. Correspondingly, this means that acquiring "general" knowledge is becoming an increasingly important part of the individual's initial education because acquiring more specialized knowledge in the course of lifelong learning activities is easier when one has a broad general education as a foundation. A sound general education is also an important prerequisite for mobility between companies, industries and occupations. Looking at vocational training, this shift in the weight assigned to various types of training content could act to dampen companies' willingness to provide vocational training. It is imperative that parallel measures be taken to avoid this.

The shifts in the importance of the general versus the specialized components of an individual's knowledge that occur over the course of a person's working life could be counteracted by re-organizing the relationship between initial vocational training and continuing training. The possibility of shortening the primary training period and using the time "saved" at some later date for periods of continuing education and training as a means of better adjusting human capital to changing training requirements should be examined. **Institutionalizing lifelong learning** in this way could provide a signal to the work force that training does **not** end upon certification or graduation.

Gearing training for highly qualified workers to international standards will make Germany's training system and companies more competitive.

Germany is known for its high level of sound, broadly-provided education. This type of education is an essential element in the German innovation system and must be preserved. On the other hand however, the notion of fostering an elite still has some negative overtones among the general public which have to be defused. A stronger **focus on producing elites** would take not only the needs of industry and government as sources of demand for highly skilled workers into account but would also make Germany more attractive to foreigners as a location for acquiring an education.

Universities in particular must adapt their structures and programs to international standards to facilitate **international recognition** of German degrees. Such steps would also make it possible for Germany to push its "education exports" as well. Besides leading to the creation of high-skill jobs, this would enhance the basis for collaboration between Germany and other countries since there is a relatively strong tendency for individuals to reestablish contact with the country they were educated in at a later point in their working lives.

People who have completed an international course of study in Germany also have better occupational opportunities and later have an easier time finding a job in a multinational company or foreign country. This also has positive consequences for the competitive strength of both the German education system and German industry: Germany profits on the one hand from such networks and the business relations based on them, while the international mobility of highly-trained individuals increases Germany's opportunities to share in the global growth of knowledge and know-how.

Information and communications technologies must be more firmly embedded in training content in order to ensure that the training received can withstand the changes of the future.

Information and communications (I&C) technologies are essential to enhancing dynamism in a knowledge-intensive society. Greater effort must be made to teach the fundamentals of information and communications technologies at all levels of education. Many service companies already lament a lack of skilled labor in general and a lack of workers with training in I&C technologies in particular. Establishing new official occupations in the information and communications field could help alleviate this situation on a medium-term basis. However, the amount of instruction content that is geared to new technologies must also be increased in the training provided for traditional occupations as well.

The threat of a shortage of I&C specialists is enormous. Signs of an engineer shortage are already beginning to emerge.

A growing number of firms, particularly technology-intensive service firms, bemoans the lack of highly trained specialists for applying and integrating I&C technologies into product and process innovations. A further shortage of specialists and engineers will put a damper on the upswing not only in the service sector, but in R&D-intensive industries as well (see Section 4.3.2.3). Despite the shortage of highly qualified I&C specialists that is beginning to emerge, there has been an only gradual increase in the number of students in these fields. More students must be induced to study science or engineering and their numbers must be kept high. Having already chosen their respective fields, today's students have determined the education patterns of new recruits into the work force up through the first years of the next century. Greater thought must be given to short-term options for mitigating the effects that the shortage of skilled personnel with I&C training will have on German companies' competitive strength and growth.

The flip side of Germany's innovation system is particularly evident in the shortage of I&C specialists: The education (and training) provided by the system tends to bind human capital to employers on a long-term basis. Further, pupils and students choose their respective careers on the basis of current hiring behavior – in other words, on the basis of cyclical trends – rather than with an eye to the long-term outlooks that a particular course of training has to offer. The German training system is not (yet) suited to modern-day demands because the trend toward specialized training continues unabated. Consequently, education policy constitutes – in a number of respects – the fundamental starting point for fostering Germany's transition to becoming an information and service society. Increased public and private investment in human capital is the key to further improving technological performance.

Education is the best insurance against unemployment.

The rapid transition to more know-how-intensive industry also has a downside because innovation has a very selective impact on the job market: Employment among the highly qualified is rising, whereas unskilled labor has seen its relative position on the job market worsen over the long run. This trend has continued on through the upswing and is even to be found in the service sector which is frequently looked to as a cure-all for alleviating employment problems. This is also the case in the USA where even a booming job market cannot obscure the growing gap between rich and poor.

Innovation leads to job cuts because it raises productivity levels. On the other hand however, without innovation, international competition would pose a direct threat to unproductive jobs. Consequently, there is no alternative to innovation-oriented policies. The job market's biggest problem is not overly qualified persons but rather those persons with little training or the wrong type of training, whose work is increasingly easy to automate. Education policy must deal with two challenges: Firstly, it must create new opportunities for people who have seen their employment options reduced as a result of structural change. Secondly, it must develop training opportunities for people who do not meet the requirements of Germany's dual education system. Education is the best insurance against unemployment. However, if structural unemployment among low-skilled persons is to be reduced, consequences will have to be drawn primarily in the labor costs area.

Put funding for human capital on par with physical capital.

Government assistance must give human capital formation and real capital formation at least equal treatment. Germany offers an abundance of financing assistance for investment projects which often favors simple givens (real investment and construction investment) and is so firmly established that it is virtually not available for other purposes (such as important regional development measures). However, favoring the one means discriminating against the other. Straight investment promotion **discriminates** against "education capital" and "technical know-how" by improving the relative price of employing real capital, compared to the price of using training and know-how. Traditional financing assistance favors "industry's core areas" which do not necessarily coincide with **innovation's core areas**.

(B) Strengthen Germany's overall R&D potential.

The quality and diversity of a country's R&D infrastructure is an important factor in competition.

Research and development enjoy comparatively high status in Germany. Highly qualified, creative people are employed in the public and private R&D field, and companies have a high level of internal know-how at their disposal. The training engineers and skilled workmen receive is considered outstanding. The **advantages** Germany offers as a location for research and development include not only industry's own extensive R&D activities, but also the country's broad science and research capabilities in the areas of basic research, applied cutting-edge research and strategic research (conducted at *Fachhochschulen* – practice-oriented technical colleges – universities and non-university research centers such as institutes belonging to the Fraunhofer Gesellschaft, Max Planck Society, the Helmholtz Association and the Blue List). Technology-oriented companies' access to technical know-how is considered to be good but could also be **expanded** (see Section 4.3.3).

Regular reviews and ongoing efforts to increase the efficiency of the government R&D infrastructure are increasingly important tasks.

Although Germany cannot allow its basic research activities to be subordinated exclusively to short-term market demands, it is necessary to inject greater responsiveness into the system. This requires recognizing signals emanating from the market at an early stage on the one hand and keeping companies apprised of research goals and progress in a timely manner on the other.

Strategic research is a **provision for the future**. When strategic research is too narrowly defined or geared to goals that are too short-term, it runs the risk of becoming inflexible. However, it is also legitimate to ask about the economic effects of basic and applied research – particularly in the non-university realm (see Section 4.3.3). Even when their differing tasks and responsibilities are factored out of the equation, Germany's individual research centers are highly heterogeneous – particularly in regard to the marketability of the research conducted by non-university facilities – which in turn necessitates a thorough re-analysis of their tasks in light of changing economic conditions. Further, more attention should be given to the **temporary nature** of government assistance in the establishment of centers of excellence. The reason: Problems arise in setting a new course and with reorientation when the emergence of new duties and responsibilities eliminates a facility's initial "mission."

Germany must continue to systematically follow the course it has chosen of increasing the competition between facilities belonging to the country's R&D infrastructure. By doing so, it would create space for enhancing the scientific quality of research being conducted and bring about an increase in the marketability of their research findings. It would however require making rigorous use of the instrument of peer review. At the same time, this would also provide a starting point for reducing the importance of basic funding for institutions which has reached a level in recent years that has increasingly circumscribed the decision-making scope of research and innovation policy.

Maintain industrial research and development at stable levels.

Encouraging industry to undertake more R&D again has to be one of the most important tasks to be tackled. Providing funds to enable small and medium-sized enterprises to **hire more R&D personnel** is a fundamentally adequate approach to this task because this would intensify these companies' know-how which would in turn increase their competitive strength over the long term as well. For this reason, steps must be taken to maintain and stabilize the involvement of small and medium-sized enterprises in industrial research (see Section 4.2.1).

In contrast to many of its European competitors, Germany no longer grants tax breaks for R&D expenditure. Tax breaks would however be easy to justify in light of R&D and innovation's external effects. Therefore, the indirect funding of R&D intensification via tax privileges or an allowance-like type of assistance would be worth considering. This would also be neutral in regard to the use of production factors, something which cannot be said of personnel transfer measures. The more indirect such assistance is, the fewer resources are necessary for distributing grants or subsidies. Today however, where public coffers are empty, attention must also be drawn to the opportunity costs of such measures. Therefore, funding measures should be designed to keep the bandwagon effects that are inherent to any type of assistance to an absolute minimum. Thought must also be given to whether the available supply of skilled labor could cover the increased demand that such a measure would trigger.

Apart from the generation of technical know-how, another decisive element of technological performance is the ability and capacity on the part of companies to adapt technical know-how produced at research facilities and to collaborate with partners within the innovation process. The **transfer of knowledge between individuals – on a "mind-to-mind" basis –** is all the more promising given that a company's ability to collaborate with research facilities and with other companies on R&D activities grows with the addition of new R&D personnel. Companies with their own R&D facilities are virtually the only ones that are in a position to make use of the opportunities offered by technology policy to expand their adaptation capacity. As a rule, small and medium-sized enterprises enlist external technical know-how to **complement** their own R&D and innovation efforts: The higher their own R&D intensity, the more know-how they absorb from external partners. It is very seldom that a company looks to collaborative activities with research facilities as a substitute for its own internal research efforts. Rather, such activities stimulate them. Opportunities to collaborate are an inducement to invest more in R&D. Even outsourcing does not generally lead to a reduction in the amount of in-house R&D a company conducts. Instead, outsourcing often goes hand in hand with an increase in the respective company's own R&D efforts.

Strengthen existing networks and innovation partnerships – set up new networks.

A know-how and technology transfer that follows the mixed top-down/bottom-up principle can be set up through an exchange of persons within industry and between the science community and industrial sectors. A wide variety of frequently proven and practiced options for collaboration is to be found between R&D-intensive industry and its sources of demand on the one hand and between industry and science on the other. Chinks cannot be allowed to develop in existing networks or **collaborative structures** as a result of essential elements dropping out of them. Conditions have to be established that make it possible to match up competences with their appropriate counterparts. The transfer of know-how functions well with regard to "insider" companies. However, there are still too many "outsiders" who work on developments of their own only from time to time and therefore have virtually no ties to government-funded R&D facilities (see Section 4.3.3).

Studies conducted in recent years clearly indicate that the boundaries between basic research, applied research and product and process development are particularly blurred in the transfer of technology between science and industry. There are a number of concrete options for reform in this area (starting with civil service regulations and extending to budgeting practices for institutions within the R&D infrastructure) which could help increase the level of personnel mobility between firms and government-funded R&D facilities and enhance opportunities for public-private partnerships. Given that Germany has so many research institutes, developing more differentiated identities for them which would adequately reflect their respective tasks and technical means also appears to be important for ensuring that interested companies find the right support.

However, industry is also called upon to take action: The "managing agents" in companies who maintain personal contacts with other companies and scientists and who are responsible for implementing R&D findings in their own companies are considered to be a decisive factor in the success of technology transfers. In-house innovation management should attach greater weight to company efforts to systematically search for know-how.

(C) Stimulate start-up activity and growth in the service sector.

Strengthen innovation activity in the service sector.

Innovation is just as important for the service sector as it is for industry. In the past, services were particularly labor-intensive. Today, their technology content is growing, particularly due to the use of information and communications technologies. As a result, technology-intensive and I&C-intensive service providers have become as important for technological performance as manufacturing firms involved in cutting-edge and advanced technologies are.

The more support the industrial base has from upstream, "accompanying" and downstream services, the more likely it is to grow (see Section 3.1 and Section 3.2). Service firms are becoming "technology providers" on the one hand, while putting pressure on manufacturing companies to produce products that meet their technical and quality standards on the other. The demands made by **lead users** from the service sector prompt the producing sector to ever-greater research achievements. Innovative "networks" are forming between industrial cutting-edge technology and high-tech and I&C-intensive service providers. Germany should strengthen its overall service sector in order to be able to improve its technological performance in high-tech industry and the service sector.

If innovation policy is to be efficient, it must also set the right priorities. It would be advisable to adjust current priorities, *inter alia*, because as society makes its way to becoming a service society, the role of primary contributor within value-added chains is shifting from hardware production to software, and – within industry – from producers to users. Although many expansive service fields are not bound to the locations where their technology is produced, they do depend on the use of advanced technical solutions to fulfill their original functions. In this connection, **funding** must pay greater attention to **potential applications** and the optimal **combination** of technologies in the service sector.

Given the fact that Germany's **research infrastructure** has done little to bring itself in line with the structural shift toward the **service sector**, it is important to set new priorities which will induce universities and research institutes – whose R&D activities are still very strongly geared to industrial production and development processes – to offer corresponding courses of study.

Support the founding of new businesses which offer new product ideas.

New start-ups and young technology-oriented companies are expected to generate impetus for employment and the use of new technologies. For this reason, the current discussion considers them to be of great economic importance for a country's innovation activity and structural change (see Section 4.2.3).

In addition, many small and medium-sized companies that conduct research offer advantages that increase Germany's attractiveness as a location for industry, R&D and innovation. Multinational corporations find good opportunities in Germany for high-quality, division-of-labor production. Together with small and medium-sized enterprises, they form corresponding producer-supplier networks which foster ties to Germany. Besides playing an important role in the diffusion of know-how and flexibly serving smaller market segments, many small and medium-sized enterprises are also highly innovative and can be found in the front lines of progress.

Particularly during the founding and start-up phase of small and medium-sized enterprises, the greater the degree of novelty is, the stronger the **head wind** they encounter and, as a result, the greater the number of projects will be that are delayed, broken off or hindered. The high level of economic risk involved in bringing new ideas in R&D-intensive industries to market maturity and the large amount of capital needed to achieve this are two factors in particular that have negative repercussions. A lack of capital affects first and foremost small, medium-sized and young enterprises (see Section 3.2 and Section 4.2.3).

The availability of venture capital for innovative start-ups has increased enormously in recent years. Government assistance measures and the expansion of exit opportunities for investors (such as with Germany's "New Market") have played a decisive role in improving this situation. Other measures that could be taken to ensure the development of Germany's venture capital market include introducing an Anglo-Saxon-style pension fund, improving the general tax rules for young, know-how-intensive companies and their investors, and granting tax privileges for investing in venture capital funds. Current efforts to reform corporate taxation and developments on the venture capital market must be taken into consideration when answering the question of how expedient these measures are. A careful evaluation of the opportunities and means offered by the individual measures would be advisable.

College graduates and people who work for government-funded research facilities often have the potential for founding their own businesses. This potential still goes largely unused in many fields. Given the fact that the number of people obtaining university degrees in scientific or technical fields is on the decline and competition over highly qualified specialists is growing, it appears that it will be necessary to mobilize this potential if Germany is to ensure and expand a sufficient supply of college graduates who are willing to start their own business. Colleges and universities could help in this effort by adding "entrepreneurship" to their instruction.

(D) Improve the institutional makeup of Germany's innovation system.

The continual enhancement and ongoing adjustment of basic conditions will continue to be core tasks for education, research and innovation policy.

If businesses are to be induced to increase their investment in production, research and development, Germany must also offer attractive market potential and production conditions for up-market products and services. Any measure that improves investment conditions and return on investment constitutes an **incentive** and is conducive to innovation. Research and innovation policy must go hand in hand with improvements in market mechanisms on Germany's product, product service and factor markets, with changes in the financial system, labor relations and labor regulations, and with reform in the education sector.

Eliminate and re-regulate impediments to innovation.

Good regulatory systems establish stable conditions and offer legal protection for bona fide acts on the one hand, while leaving sufficient leeway for economic-technical creativity on the other. This is not the case in all areas in Germany. Although a high degree of regulation is typical of highly developed economies, Germany should take steps to prevent over-regulation. It should also take care not to allow its regulatory system to ossify.

All conditions that are relevant to innovation must be examined closely to determine whether they constitute an impetus or obstacle to innovation.

Innovation policy must take on a broader scope and be interleaved with other political fields that are relevant to innovation. A number of impediments to innovation are fundamental in nature and can have a prohibitive effect (such as high market risk or the lack of demand for new, advanced goods and services).

- Many companies call for faster **approval procedures** for investment in new plant and equipment or in facility expansion. They also lament that requirements and approval procedures for investment and new production processes are often not coordinated, making the outcome virtually impossible to predict. Another point cited is the possibility of retrofitting requirements being imposed on already approved facilities. Despite these complaints, the quality of the permits being issued should not be sacrificed for the sake of faster processing. Approval procedures must be organized on the basis of clear-cut parameters and be coordinated by the offices involved in issuing the respective type of permit. Measures taken to speed up the investment approval procedure have already served to build confidence. Other improvements would include the establishment of central points of contact for all regulatory requirements; such offices could assist companies wanting to, for example, build new facilities or modify existing facilities find their way through the country's manifold regulatory landscape.

- Product **approval procedures** are time-consuming, have a negative effect on motivation, and depress the profit to be made from innovation. It is however impossible to foresee market volume and profits in advance, particularly when it comes to new technologies. Industry that conducts research needs **parameters** that remain **calculable** on a long-term basis. This impediment sometimes works to deter German innovators who often lead the way in their markets. It is also extremely important by international comparison. The situation in Germany has improved somewhat thanks to a number of factors such as the new Genetic Engineering Act which enables genetic engineering research and biotechnical production in Germany. Further progress is necessary. Efficient coordination within Europe is also particularly important if the advantages offered by the Single European Market are to bear fruit. Changes in approval procedures have an enormous impact on innovation activity. This was evidenced most recently by the harmonization of licensing practices in the European Union's health care sector. Successful regulation will stimulate innovation activity in Germany only when the complementary resources needed for making use of newly won latitude are also available. This is illustrated by Germany's competitive strength in obtaining licenses for new medicines – which cannot be brought fully to bear at present because of a lack of research orientation and specially trained personnel in the country's university clinics.

It is no accident that Germany's service sector exhibits less dynamism than the service sectors of other highly developed economies do. Many areas of Germany's service sector have long histories as regulated, protected fields with little competition. These institutional and legal conditions are part of the country's "national innovation system" and have to be "re-regulated." Reviewing and pruning government regulations could be of great help in getting Germany's supply of services onto the fast track and boosting the industrial sector's innovation intensity (see Section 3.3).

Expand the protection of intellectual property.

Patent protection and the protection of intellectual property are becoming noticeably more important in the wake of globalization. This issue also assumes a new quality in view of the growing importance of services and international information networks (see Section 3.2). German companies are particularly dependent on the protection of their inventions because they are frequently international front-runners. Technical know-how is becoming increasingly codified with the spread of modern I&C technology, which makes it also easier for others to copy it. The possibility of being able to protect technical know-how against imitation and thus generate revenue from innovation constitutes a key incentive for investment to expand technical know-how. In light of the structural shift toward innovation-oriented services, patents are not always able to provide this protection. As a result, **other types of protective rights** are becoming increasingly important. Even so, great value should continue to be attached to the public relations work done to educate and inform inventors and small and medium-sized enterprises about the European and international patent application systems (which have progressively developed in recent years). Thought should be given to introducing grace periods.

Promote greater acceptance of technology.

Parts of the population have a contradictory attitude toward technology. While individuals make use of R&D-intensive products such as pharmaceuticals, efforts to establish new production locations for such products often meet with opposition. Problems with public acceptance are not limited to the production of new technologies but also extend to the demand for them (such as the use of genetic engineering for agriculture). This lack of acceptance of new technologies could be mitigated were the political sector to credibly prove its commitment to new technological trends, their problem-solving ability and the financial opportunities they entail, and were it also to disclose their impact at an early stage. To facilitate an assessment of the risks involved in new technologies, comparisons should be made with risks in other areas that are already familiar to the general public, and efforts to communicate the opportunities these technologies offer should be stepped up. This task is not just the responsibility of the political sector but also of the other players in the innovation system, particularly in the science field.

Establish clear-cut responsibilities – eliminate mixed financing.

As a rule, the principle of subsidiarity should be observed when assigning **roles** to various local and regional authorities and to areas of **purview** in research and innovation policy on the one hand and to departmental provinces that are relevant to technological performance on the other. It would be useful to assign **strategic** tasks and the setting of political priorities in interdepartmental initiatives they necessitate to the **federal** level. These would include strengthening the venture capital market and the opportunities and means for starting a business, and establishing and safeguarding competitive structures (such as telecommunications, the energy supply, banking), and the search for solutions to societal problems in the health care, nutrition, communications, transport, energy, environmental protection and aerospace fields, for example. If necessary, the **European** level could be drawn upon when seeking especially demanding, cross-border, large-scale technological solutions (such as space travel, defense, energy).

This tack should be held to in order not to waste time and energy. However, **mixed financing** and the unclear division of labor between various levels present problems time and again, particularly in connection with the "assignment problem" – namely, both in the relationship between Germany's federal and *Länder* governments and in the relationship between Germany's funding policy and supranational organizations. In the case of the European Union, these two factors lead to unbridled breaches of the subsidiarity principle because EU assistance meanwhile intervenes with the transfer of technology and small businesses policy with an inflationary number of programs which are outfitted with only minimal funding.

Mixed financing also leads to the federal and *Länder* governments blockading one another. One particular effect of this situation is unclear priorities. However, with money currently in short supply in the government's coffers, it is particularly important to rank priorities so that funds can be freed up for projects that have been assigned higher priority. Mixed financing can prove to be a major hindrance in such cases. This is particularly true of jointly provided (i.e., by the federal and *Länder* governments) basic funding for institutions because regional interests also play a role in each and every decision.

Another critical negative effect had by mixed federal-*Länder* financing is the fact that it **effectively discriminates** against indirect-specific, indirect and direct funding for research. The reason: Those major research institutes which largely receive joint basic funding are bridled by their personnel expenditure which can be reduced only with difficulty. Basic funding for institutions has therefore tended to expand as a result of the institutions' own inertia. On the other hand however, project funding and indirect-specific funding have served as "quarries" in times of financial constraint and have been used to fill in gaps where necessary. Direct and indirect-specific funding measures have eroded. This has led to a level of fragmentation that leaves these measures virtually indiscernible at any level.

Recourse to EU funds offered a (possibly welcome?) way out of this situation – but also led to flagrant breaches of the subsidiarity principle. In this regard, the use of mixed funding to provide basic research funding for major facilities that received basic funding was one of the reasons why direct/indirect and transfer-oriented measures have shifted to the EU.

(E) Education, research and innovation policies are an integral component of a more broadly defined economic and sociopolitical agenda.

Nearly all political departments influence innovation activity in the industrial sector: For example, health policy has a decisive impact on innovation activity on the part of the pharmaceutical industry and suppliers of medical equipment, whereas the decision-making domain of the Federal Ministry of Economics influences the development of innovative (public) transport services. In addition, the analysis of the national innovation system clearly reveals the diversity of the participants in the system. There is however a constant danger that coordination sometimes misses the mark as a result of this diversity. Looking at Germany's education, research and innovation policies, this poses the task of developing – in conjunction with the various players involved – abstract **objectives** into concrete, marketable solutions, albeit without prescribing the technological parameters to be used or their translation into product and process innovation. These objectives could provide the focal point on which the coordination of the individual decisions of all those involved in the process centers. This would require the political sector to credibly prove its commitment to new technological trends, their problem-solving ability and the financial opportunities they entail – without veiling potential inherent risks. In this sense, education, research and innovation policies are both a **cross-sectoral and a management task** which plays the role of advocate for innovative solutions, expanding these policies beyond their traditional responsibilities.

Part IV   Measurement and Interpretation of Germany's Technological Performance 1998

1   Basic Concepts

1.1   Indicators and the innovation system

An economy's technological performance is based on a number of factors. The ability to transform new ideas, products and processes into economic wealth and social prosperity is of central importance in this regard. The ability and willingness of firms to pursue innovation, the conditions underlying the implementation and dissemination of new technologies, and the existence of a broad knowledge base within an economy play as much a role as do macroeconomic developments and general framework conditions. This large number of qualifiers also explains why it is not possible to draw a picture of an economy's technological performance **using just one "point of departure."** Rather, a variety of indicators must be used to depict the development of those parts of an economy that are pertinent to an assessment of technological performance. Only when these indicators are viewed on an overall basis is it possible to use them to describe and assess the current situation and prospective trends.

Another important reason for taking this approach is the fact that economic research has yet to develop a self-contained innovation theory that could also be applied at macroeconomic level, thereby enabling a straightforward assessment of an economy's technological performance. For this reason, several explanatory approaches from the economic and sociological fields are taken into consideration here for interpreting and assessing the innovation processes taking place within a particular economy: This report draws upon approaches from neoclassical innovation economics and institutional economics, considerations from the new growth theory, and sociological approaches and puts on them equal footing with one another, viewing them as complementary rather than substitutive explanations for actual innovation processes.

Innovation as an intermeshed process

As a rule, the individual steps in a company's internal innovation process (market analysis, research and development, product and process design, production, and marketing) are closely interwoven and mutually dependent. Building on what is learned in previous phases, these steps are repeated in ensuing phases of the innovation process. According to this picture of the innovation process, the development of innovation entails long-term interaction between all involved. Consequently, the innovation process should not be interpreted as being simply a process in which technology is passed on to the next player, much like a baton in a relay race. The production, application and implementation of know-how are closely intertwined and therefore very difficult to separate when transferring know-how to the next player in line or to the next phase in the process. The scenario in which a "technology push" occurs and is followed by a more or less automatic innovation process rarely reflects reality. Successful product and process developments are just as dependent upon the overlaps and feedback occurring within this system as they are upon docking with external information flows.[4]

When examining an interdependent economic or innovation system, it would be sensible to use indicators for the various areas under consideration as an avenue for accessing performance factors. Typically, close intermeshment, broad feedback, and cross-fertilization are critical to successful innovation. This report on Germany's technological performance takes its bearings from this interactive method.

This type of approach to the innovation process is outlined in Chart 1.1, starting with potential factors of production and then proceeding to the development and implementation of know-how, all the way to fruition and success. This chart can be used to shed light on the various "stations" in the innovation process, elucidate macroeconomic conditions, outline criteria for success, and present the respective indicators used to measure technological performance:

- Know-how arises out of innovative processes. **Research and development (R&D)** and the level of **vocational qualification offered by the working population** are vital inputs for these processes. R&D staffing and expenditure levels normally serve as indicators for research activity. They represent front-end investment which should subsequently amortize itself in the course of the following years. They also provide information about intersectoral and intra-industry structural change, the degree of orientation toward promising markets and the intensity of renewal processes within a particular company.

---

[4]   S. Kline and N. Rosenberg (1986), An Overview of Innovation. in: R. Landau und N. Rosenberg (Eds.), The Positive Sum Strategy: Harnessing Technology for Economic Growth.

- Human capital and the ongoing expansion of technological means being fuelled by outstanding scientific achievement provide the foundation for innovative processes.

- Education and science constitute the fundamental **potential factors of production**. Using statistics on expenditure on education, education efforts can be documented, attributed to the respective agency and evaluated on an international plane. The procedure is similar when evaluating the cost of the dual vocational training system to industry.

- Education patterns among the work force (such as the number of people with a degree in science or engineering) and figures on the level of participation in continuing education and training measures provide insight into the present and future know-how base to be found among companies in Germany and consequently into the opportunities and risks arising for Germany's level of technological performance.

- Publication rates can be evaluated to determine the international standing of a country's scientific field. Patent applications submitted by scientists provide evidence for the marketability of scientific efforts.

Chart 1.1   The innovation process and indicators of technological performance

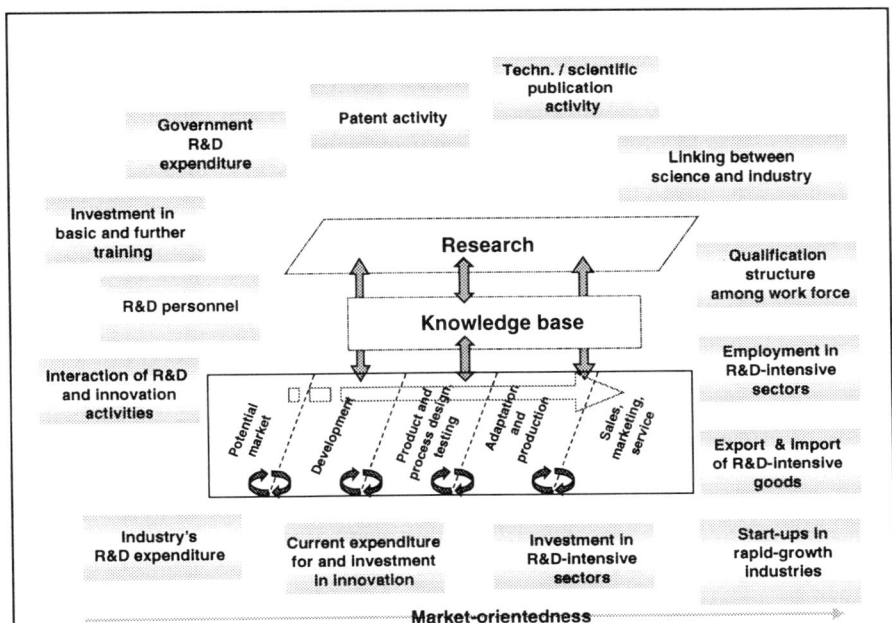

Source: Own Chart

- A comprehensive picture of **innovation activity being conducted by private business** that goes beyond R&D can be obtained by evaluating various innovation indicators. These indicators include innovation expenditure, internationalization strategies and innovation goals, barriers and hurdles in the innovation process, and the amount of export volume and share of turnover attributable to new products. Networks between companies and between industry and science can be ascertained by analyzing know-how flows, transfer mechanisms and collaboration partners and, in doing so, by pointing out differences relative to size, sector and industry.

- Patents are the product of research and development conducted by companies. Consequently, **patent indicators** draw a more accurate picture of the marketplace than other indicators do. Besides providing initial indications of how new know-how could be marketed, they also point to patterns of technological specialization and show the commercial potential to be found in – for example – "key technologies."

- Global trade shares and **export-import ratios** are generally used to assess relative "strengths" and "weaknesses" on an international plane. They augment output figures and employment indicators by depicting the structure of visible trade, outlining comparative advantages in the global marketplace and bringing global markets into the equation.

- In addition to the level of investment in the modernization and expansion of production facilities, the number of **new start-ups** can also be used as an indicator of ability to undertake structural change. Using this indicator, it is also possible to estimate the speed at which intersectoral and intra-industry structural change that comes "from below" is progressing. Start-ups in know-how-intensive areas of the industrial and service sectors are particularly important in this connection.

The approach chosen for this report on Germany's technological performance endeavors to elucidate the "ground rules" that govern the innovation system and shape its individual elements and their interaction. This approach aims at providing answers to questions about systemic relations and at offering not only conceptual considerations but also information which can be used to make viable empirical statements regarding the level and development of German industry's technological performance. In order to interpret the facts, it is necessary to draw comparisons with other economies and conduct an analysis of change over the course of several years. The objective of this report is to examine the importance of the system's individual elements and draw attention to correlations within the system.

1.2 National innovation systems and their institutional framework – an introduction

Over the long term, national differences will shrink and competition over internationally mobile resources will pick up as a result of the progressive liberalization of commodity and capital markets and the growing means of modern communication. **Paradoxically**, the more similar national economies become, the more important other differences will be for growth and prosperity. For this reason, the peculiarities of a respective country's economic structure will become increasingly important in a globalized world – and in the age of a single European currency. This is particularly true of those elements of a country's economic system that shape its ability to develop innovation and are the subject of the debate over "national innovation systems."

This debate is aimed at obtaining a better understanding of the overall context behind the generation, diffusion, adaptation and exploitation of new know-how.[5] The following sections outline considerations regarding "national innovation systems" and seek out approaches for gradually expanding this report's coverage in such a way that it is possible to progressively develop a better medium and long-range picture of the factors determining Germany's technological performance. A practicable method has however **yet to be found**.

1.2.1 National innovation systems

At national level, the innovation activity conducted by individual firms is embedded in an extensive meshwork of incentives, rules, institutions and regulatory structures. Depending upon the form they take, these conditions have a significant influence on the intensity and direction of corporate innovation efforts.[6] This is because a number of factors play an important role in the functioning of an innovation system and in the level of corporate willingness to pursue innovation in a knowledge-based economy. These factors include technological means and human capital, mechanisms to protect revenue generated

---

5   Cf. R.R. Nelson and S.G. Winter (1982), An Evolutionary Theory of Economic Change. When examined from the standpoint of institutional economics, differences in national innovation patterns can be attributed to systematic differences in the respective institution-related incentive structures. This approach supplements but does not supplant the neoclassical innovation theory with the theoretical framework of the "new economy of organizations." For more on this, see P. Milgrom and J. Roberts (1992), Economics, Organisation and Management, and O.E. Williamson (1985), The Economic Institutions of Capitalism.

6   Cf. also OECD (1997), OECD Proposed Guidelines For Collecting and Interpreting Technological Innovation Data - OSLO Manual; OECD (1998), Technology, Productivity, and Job Creation – Best Policy Practices. Highlights.

by innovation activities, know-how and its transfer.[7] The degree to which **incentives to foster innovation** are linked with one another is of fundamental importance to the system's ability to function.

The peculiarities exhibited by a particular country's innovation system are not necessarily the result of planned, systematic considerations. Rather, these differences have developed over time and might even have been caused by individual events or produced by coincidences of history. Furthermore, national innovation systems are subject to gradual but continual change. The driving forces behind this change are the exchange with other innovation systems, the **globalization of innovation activity** – particularly on the part of major corporations – and, in Europe, the growing impact of European integration arising from the harmonization of regulatory systems and the European Union's increasing involvement in the technology policy field. And finally, a country's economic and technology policies also generate impetus for change in its national innovation system. One example of this is the deregulation of the telecommunications field, its impact on telecommunications service providers and the resultant innovation activity this triggered in other parts of the economy. Another example is the (worldwide) reduction in government spending on armament research, which has led in part to a fundamental realignment in many high-tech fields.

What is a national innovation system?

An innovation system encompasses much more than just a country's national research infrastructure and the research efforts undertaken by industry and service providers. Although particular attention is paid to universities, research institutes and technology transfer systems when considering innovation systems, a number of other factors are also decisive to the system's functioning – in other words, to its technological and economic success. These factors include human capital, basic and further training, the infrastructure, the credit and capital markets and the attendant system of corporate governance, government regulation, state demand for innovative goods and services and, last but not least, market potential, exchange rates and formal and informal communication mechanisms as well as networking between companies and between companies and the research infrastructure.

---

[7]   For an overview, see: P. Stoneman (1995), Handbook of the Economics of Innovation and Technological Change; M. Dodgson and R. Rothwell (1994), The Handbook of Industrial Innovation; Ch. Freeman (1994), The Economics of Technical Change; G. Dosi (1988), Sources, Procedures, and Microeconomic Effects of Innovation; G. Grossman and E. Helpman (1993), Innovation and Growth in the Global Economy; P. Krugman (1991), Geography and Trade; G. Dosi, K. Pavitt and L. Soete (1990), The Economics of Technical Change and International Trade.

Differences in the technological means and market conditions in an economy's individual sectors, *inter alia*, give rise to sector-specific features in the respective national innovation system. These specific features include in particular the type of market competition, the opportunities available for collaboration with other companies, the occupational and entrepreneurial incentive systems offered in the science and research fields, the transfer of knowledge and know-how from universities and research institutes to business, and the criteria for the development and establishment of technological and social norms and standards.

Consequently, the political aspects of an innovation system include not only research and innovation policy but other political fields (such as finance or the labor market) and the institutions operating in them. And finally, this list must also include the defense, health, environmental protection and transport fields as well as communications and foreign trade relations which are particularly material to R&D and innovation decisions. The system is also augmented by semi-private and private institutions such as chambers and associations.

Although a number of economies exhibit a comparable "innovation potential," their areas of technological focus have developed differently over the years. Technological progress is "path dependent." In other words: traditions and long-term factors lay the foundation for the respective innovation system and its development. Key institutions for company-level innovation capability such as the education and science systems, the system for regulating the labor market, and the corporate financing system exhibit strong national peculiarities, and the information at hand does not indicate that this will change much in the future.

1.2.2 Distinctive characteristics of the German innovation system

The principal features of the German innovation system are outlined in this section to facilitate understanding of the empirical facts described in subsequent sections.

Institutional traits in comparison

The strengths and weaknesses of German innovation activity are due to a weave of dependencies which is specific to the German economy. These dependencies exist between companies, customers, employees and shareholders; they are organized and structured according to institutionalized rules within a framework of incentives and restrictions. The institutional structure is to be found primarily at national level. Regional and sectoral characteristics that are specific to Germany do not deviate from the country's general pattern. In Germany, the institutional structure fosters the development of long-term, collaborative, consensus-oriented relations – between companies, between companies and their employees, and between companies and their owners. Also to be taken into account is the fact that this entails specific advantages that go hand in hand with specific disadvantages (in other words: trade-offs).[8]

The following factors shed light on the peculiarities of the German innovation system:

- A **corporate financing system** in which banks and long-term commitment on the part of important shareholders play a strong role, making long-term financing possible. However, this long-term involvement often restricts the speed at which a changeover into new markets or technologies can be made.

- A **system of industrial relations** in which trade unions play an important role. This system enables collaborative relations within companies and makes it possible to coordinate wage negotiations so that they cover more than just individual companies. The disadvantage to this system is that it interferes with a company's ability to abandon long-standing markets and manufacturing activities in order to tap new markets.

- An **education and vocational training system** which fosters thorough training in preparation of employment and enjoys strong involvement on the part of industrial associations, companies, unions and management. Since these investments in human capital are highly tailored to the particular company and industry involved, they reduce the degree of the labor force's mobility between companies, occupations and industries.

- A **system of intercompany relations** which enables close collaboration between companies in the areas of technology transfers and standardization. On the other hand however, establishing standards by consensus makes it harder for market-oriented standards to assert themselves via the innovation arena.

- An **innovation policy** which, given Germany's strong regional and institutional diversity, provides greater support for the diffusion and adaptation of new technologies than is found in other countries.

The following example serves to illustrate the interdependencies within this system: An effective **vocational training system** in which a company is willing to invest requires, *inter alia*,

- options for long-term financing (because this investment pays off only in the long term),

---

8   Cf. in this regard the extensive description of trade-offs in: CPB Netherlands Bureau for Economic Policy Analysis (1997), Challenging Neigbours. Rethinking German and Dutch Economic Institutions.

- coordinated wage agreements (to reduce the risk of "poaching" and, as a result, "free riders" as much as possible), and

- collaboration between companies in the area of technology development and standardization (to ensure that skills for which examinations are held and certificates are issued can be compared).

Institutional traits are of importance in the international contest over innovation and for technological performance because they give rise to incentives or restrictions for various interest groups within the respective company and along its periphery (employees, owners, managers, investors, etc.). In addition, there are important **intermeshing complementarities** between institutional traits. In other words, the effect these traits have is not just cumulative in nature but also mutually reinforced. Consequently, the absence of one element impairs the working order of not just the other individual elements, but of the entire system as well.

Table 1.1:   Comparison of institutional frameworks

	Consensus-based market system	**Liberal market system**
Industrial relations		
Wage negotiations and relations between employers and employees	Employers' associations and trade unions play an important role in the formal and informal coordination of wage negotiations and relations at sector level. The degree of mobility observed among skilled workers is relatively small; wages and wage structures exhibit little flexibility.	Negotiations are largely held on a company-to-company basis and are not coordinated. High degree of mobility among skilled workers; very flexible wages and wage structures.
In-company industrial relations	The bodies elected by employees play an important role in company decision-making; they have ties to non-industry unions; they also enjoy representation on supervisory boards.	Worker representatives remain in the background during collective wage negotiations. They have little say regarding the terms of employment contracts in the private sector. Unions play a relatively small role.

Education and vocational training		
Vocational training	Vocational training is very important; industrial organizations and trade unions are extensively involved in the system.	Workers at lower levels seldom undergo continuing education and training. Subsequent compulsory vocational training is important
University system (engineering in particular)	Training for engineers is closely linked to industrial technologies; professional associations are strongly involved. There are also doctoral programs in basic sciences and in engineering which have close ties to large companies.	Training for engineers is not closely tied to any specific technology. Doctoral programs in basic sciences and in engineering do not have close ties to companies.
Company financing		
	Publicly quoted companies have stable shareholder systems. Banks play an important monitoring role. Similarly, opinions from various sides are obtained for monitoring purposes. Hostile takeovers are difficult.	The existing regulatory framework allows the hostile takeover of companies that are listed on the stock exchange. Venture capital is available for high-risk projects.
Inter-company relations		
Method of setting standards	Standards are set on a consensual basis within the respective industry.	Standards are set on a market basis.
Competition policy	Very intense competition in foreign trade whereby direct competition is avoided via product differentiation. Business connections play a role in conflicts over contractual relationships and when setting rules for the regime.	Competition policy attempts to prevent collusive or concerted action. Limited regime for handling problems with contractual relationships.
Role of government / Role of innovation policy		
Public R&D infrastructure policy and the transfer of technology	Universities and research institutes have close ties to companies in established technologies. Experience is important for collaboration. R&D facilities exhibit high level of regional diversity combined with low level of regional concentration.	Limited institutional framework for technology diffusion. Strong regional concentration of government-funded R&D institutions.

Innovation and technology policy	Comparatively high level of technology-oriented assistance and support "across a broad front."	Geared to large-scale (military) projects. Concentration on individual technologies.
Regulatory framework / Regulation	Extensive regulatory framework; high degree of regulation.	General regulatory framework; little regulation.

Source: Based on Soskice, D. (1997), Divergent Production Regimes.

An understanding of the complementarities within the system is essential to being able to gauge the impact that a change in individual elements of the system would have on the working order of the entire system. For example, making in-plant vocational training more general in nature (i.e., reducing its company-specific or industry-specific focus) in order to increase the occupational mobility of people receiving this training lessens the incentive for a company to offer in-plant vocational training and, in the process, diminishes the innovation capacities arising from the present system. Accordingly, in order for measures aimed at changing the system to be as effective as possible, due consideration must also be given to their impact on the system's other elements.

Training, the transfer of technology, and technical standardization take place for the most part within organizational structures that are specific to the particular sector. In Germany, many of the primary players involved in these activities are to be found at **sectoral level**. Industrial organizations, business associations and industry-specific unions (and in some cases, chambers of industry, chambers of commerce or chambers of skilled crafts) play an important role. Employee representation in individual companies has strong ties to the unions operating in the respective sector, whereas top managers maintain close relations with industrial associations and employers' associations. Sectoral coordination is interlaced into the national institutional framework. Vocational training and employee representation are regulated by **framework legislation** which is aimed at establishing standard cross-industry practices; this legislation has the support of the country's labor courts and various expert committees.

Germany exhibits a number of institutional traits that deviate substantially from those found in the USA and Great Britain (for which empirical studies exist). As a result, Germany's innovative strengths are configured differently. These strengths entail high-quality products in established industrial sectors[9] whose production requires complex processes, maintenance service and close customer relations –

---

9   Cf., for example, the reports on Germany's Technological Performance from last year and previous years.

so-called diversified high-quality production.[10] Germany has marked strengths in "cumulative" product and process innovation in "established" technologies, most notably in the mechanical engineering, automobile, electrical engineering and chemical products fields. Manufacturers of such products require qualified, experienced workers who can be assigned a high degree of responsibility. The type of innovation that Germany produces – high-quality innovation that follows well defined trends and creates high value in established industries – corresponds to a number of important institutional conditions: availability of long-term capital, cooperative trade unions, influential employers' associations, smooth-running training systems, and close, long-term collaboration between companies and between companies, research institutes and universities.

In contrast to Germany, the USA and Great Britain are not particularly successful on international markets in these areas. Institutional factors play a large role in this connection. Germany's institutional system differs in part substantially from the more deregulated, market-oriented institutions to be found in the USA and Great Britain. German institutions foster long-term time horizons and collaboration – the prerequisites for producing high-quality goods and services which require qualified workers with good company-specific know-how on the one hand and close, long-term relations between companies on the other. By comparison, the Anglo-American institutional framework is better suited to other production strategies. The USA (and Great Britain to some extent as well) is particularly strong in "radical" innovation in the area of emerging new technologies (such as biotechnology and microprocessors) and in large, complex technical systems in exceptionally R&D-intensive fields which involve complex systems or entail giving managers substantial discretionary powers under the condition that they meet tough financial targets. Examples of this include new consumer media, defense systems, major software and computer products, and the aerospace industry.[11]

---

[10]   W. Streeck (1991), On the Institutional Conditions of Diversified Quality Production, in: E. Matzner and W. Streeck (Eds.) The Socio-Economics of Production and Employment, and H. Kern and M. Schumann (1984), Ende der Arbeitsteilung?

[11]   The two adjectives **cumulative** and **radical** which are used here for describing innovation seem to suggest associations with the terms advanced technology and cutting-edge technology. However, radical innovation, for example, can take place in the advanced technology field and cumulative innovation can be found in the cutting-edge technology field. In contrast to the differentiation between cumulative and radical which is used here basically for illustrative purposes, the areas "advanced technology" and "cutting-edge technology" are clearly defined in this report. According to the NIW/ISI list (List A.1 in the Appendix) used here, **cutting-edge technology** entails goods whose R&D intensity is more than 8.5% of turnover. **Advanced technology** covers goods whose R&D intensity is more than 3.5% but less than 8.5%.

The effect of institutional elements on innovation behavior

Despite the fact that it has chosen a different route than those taken by highly developed rival economies, Germany puts in an impressive **macroeconomic performance** (Table 1.2).

An **international comparison** of production, demand and employment structures reveals that Germany enjoys competitive advantages in the area of "advanced technologies" in particular. Although this area invests an above-average share of turnover in R&D, its research activities do not match the extremely R&D-intensive level found in the cutting-edge technology field. Nowhere else in the world are the production and employment shares attributable to the advanced technology field larger than in Germany.[12] Looking at the cutting-edge technology field, Germany reports the largest share of employment (giving it top ranking with Japan), the third largest share of value added – following Japan and almost on par with the USA – and the largest share of domestic demand. Therefore, the crucial difference between Germany and the USA and Great Britain is **not** that its cutting-edge technology sector is less pronounced but, rather, that it has a smaller-scale service sector.[13] It would be possible to conclude from this that the elements of national innovation systems presented here offer, first and foremost, rudimentary explanations for **intersectoral and intra-industrial specialization**, albeit without permitting any systematic insight into the structure-impacting driving forces at work in the relationship between industry and the service sector.[14]

The German system is apparently better suited to quickly absorbing and translating new technologies and products – regardless of whether they are from Germany or abroad – into value added and, in the process, achieving a higher degree of productivity than the USA and (especially) Great Britain[15] where top

---

[12] Cf. the reports from last year and previous years.

[13] Cf. Section 3 and, regarding the problems involved in conducting international comparisons in the service sector: DIW (1998), Das Dienstleistungs-Puzzle. Ein aktualisierter deutsch-amerikanischer Vergleich.

[14] To be in a position to elaborate on the differences that various countries exhibit within their own respective service sector and/or in the relationship between their industrial and service sectors, our examination of national innovation systems would also have to, *inter alia*, deal intensively with the issues of varying forms of labor market regulation or preferences in employment behavior. Adding these facets to the approach used here is theoretically possible and would constitute a starting point for more extensive work.

[15] The national innovation system's ability to use its broad domestic supply of skilled labor to profit from the international exchange of know-how has also been observed in Japan and various newly industrialized countries in Southeast Asia. Cf. D.C. Mowery and J. Oxley (1997), Inward Technology Transfer and Competitiveness: The Role of National Innovation Systems, in: D. Archibugi and J. Michie (Ed.), Technology, Globalisation and Economic Performance.

Measurement and Interpretation of Germany's Technological Performance

technological performance is admittedly more radical and the first rapid steps toward commercialization are also seen more frequently.

Table 1.2:   Domestic production, domestic demand, and employment in R&D-intensive sectors in selected OECD countries 1993 – 1994/1995

Sector	Former West Germany	USA	Japan	France	Italy	Great Britain
Percentage of gross value added **1993-1995**						
R&D-intensive sectors	12.2	8.5	11.5	7.7	6.4	8.0
–Cutting-edge technology	3.5	3.6	3.9	2.6	1.9	2.9
–Advanced technology	8.7	4.9	7.7	5.1	4.5	5.2
Non-R&D-intensive sectors	13.7	9.5	13.5	11.7	13.9	10.2
Manufacturing industry	25.9	18.0	25.0	19.3	20.4	18.3
Percentage of domestic demand[1] **1993-1994**						
R&D-intensive sectors	7.2	9.0	6.6	6.7	5.6	8.6
–Cutting-edge technology	4.0	3.7	2.2	2.6	2.5	3.0
–Advanced technology	3.3	5.3	4.3	4.1	3.1	5.6
Non-R&D-intensive sectors	16.2	11.0	14.6	12.1	10.8	12.1
Manufacturing industry	23.4	19.9	21.2	18.8	16.4	20.8
Percentage of employment 1993-1995*						
R&D-intensive sectors	12.8	5.9	9.3	7.5	5.8	8.2
–Cutting-edge technology	3.1	2.5	3.1	2.3	1.4	2.7
–Advanced technology	9.7	3.4	6.2	5.2	4.3	5.4
Non-R&D-intensive sectors	14.9	9.7	13.8	11.0	14.7	11.4
Manufacturing industry	27.7	15.5	23.2	18.5	20.4	19.6

[1] Gross valued added in the respective branch plus net imports and minus net exports in percent of total domestic demand (private and government consumption and capital investments). Net exports and imports were estimated with the help of the gross value added share of the respective domestic production.

* USA: 1993-1994; Great Britain:1993

Cutting-edge technology: Pharmaceutical products, EDP equipment/office machines; radio/TV/telecommunications; aerospace; precision instruments; precision goods, watches & clocks.

Advanced technology: Other chemical products; mechanical engineering products; electrical engineering products w/o radio/TV/telecomm-unications; railroad vehicles; motor vehicles.

Sources: OECD. STAN Database Economic Outlook; DIW calculations and estimates.

On the other hand however, Germany outperforms Great Britain and the USA in the diffusion, adaptation and economic exploitation of new technologies across the entire industrial base.

The crucial question is, however, **whether** the German system will have sufficiently long time frames in the future for translating innovation into value added – as product life cycles become continually shorter and the world becomes increasingly globalized.

1.2.3   Questions yet to be answered

The above observations on national innovation systems provide rudimentary explanations for differences in the technological performance of rival countries. The comments made here are neither complete – and lay no claim to having taken all facets of a particular institutional structure into consideration – nor are they static. In other words: the tiers taken into consideration here are subject to ongoing change.

Germany cannot avoid global trends: Answers must be found to the trend toward ever higher levels of know-how in industry, the rapid development of services and the growing international intermeshment arising from globalization. The importance of internationally available know-how is increasing and, as a result, so is the necessity of maintaining reserve capacity for absorbing this know-how. The tendency to recruit and hire highly qualified skilled labor from abroad will also grow. Many companies already foster a strong global investment culture which incorporates profitability considerations and market opportunities regardless of where they are discovered into their calculations. Globalization increases competitive pressure on the home front as well. This is particularly true of those sectors which have been heavily regulated to date.

The extent to which this may result in a tendency toward a gradual **approximation** of national innovation systems is unclear. It appears that distinctive national traits such as a country's particular education and training system or its research scene will not be forfeiting their role as critical determinants of international competitive strength in the foreseeable future. In many cases, regional and local players are even becoming increasingly important. Furthermore, the types of conditions under which innovation is developed and the advantages they offer that are specific to a particular country are currently taking on even greater importance due to the growing international movement of goods and capital.[16] At the same time, international companies are making increased use of

---

[16]   Cf. for more details M.E. Porter (1990), The Competitive Advantage of Nations; Ch. Freeman (1997), The "National System of Innovation" in Historical Perspective, in: D. Archibugi and J. Michie (Ed.), Technology, Globalisation and Economic Performance.

their opportunities to become involved in international financial markets. This could result in Germany aligning itself with other countries' practices which would particularly promote a culture of "hostile" corporate takeovers in Germany as well.

For this reason, monitoring the various trends and measuring change on the basis of indicators will present a challenge when analyzing Germany's innovation system for future reports. Irrespective of these future tasks, it should be noted here that in order to evaluate the various facets of Germany's technological performance it is also necessary to examine institutional frameworks as well – at regional, national and global level.

2 Industrial Specialization in Germany

Taking a longer-term view, the questions arise: What is the composition of Germany's industrial "portfolio" of goods and know-how? And: How do the focal areas pursued by German research and industry measure up internationally in terms of ensuring that Germany will continue to be capable of a high level of technological performance in the future as well? An economy's endowment with production factors determines its particular specialization patterns. As do the means which the respective country's institutional framework (regulations, organizations, division of labor, legal framework, system of public finance, system of government, resources, etc.) and societal traditions (in the form of its "national innovation system") offer. This is why economies that are similarly endowed with, for example, innovation capabilities may take different routes and differ substantially in their innovation behavior and still have comparable macroeconomic track records.

2.1 Germany's industrial structure in international comparison

The following section examines German industry's specialization patterns as reflected by the level of research intensity the respective range of products exhibits. R&D-intensive industries[17] are incubators for new technologies. Such industries bundle capabilities and capacities for invention, for the development of

---

[17] According to the NIW/ISI list of **R&D-intensive** goods used for this report, **cutting-edge technology** entails goods whose R&D share is more than 8.5% of turnover. **Advanced technology** covers goods whose R&D share is more than 3.5% but less than 8.5%. Together, these two groups comprise industry's R&D-intensive sector. **By no means** is this differentiation to be understood as a judgment that the advanced technology field should be labeled as "older" or "less valuable" and cutting-edge technology as "new," "modern" or "more valuable." Rather, these two groups differ significantly in their level of R&D intensity on the one hand and in the degree of protection they are granted on the other. Cutting-edge technology goods have the highest R&D intensity and are frequently subject to government intervention in the form of subsidies, government demand and/or protection against imports. The cutting-edge technology sector in all leading industrialized countries draws special attention from **government agencies** which use their funding to achieve not only technological goals but to pursue independent government objectives (external security, health care, etc.) in many cases as well. Cf. H. Grupp, H. Legler, Innovationspotential und Hochtechnologie. Report by FhG-ISI, NIW and Gewiplan for the Federal Minister of Research and Technology, Karlsruhe and Hanover, 1991. Cf. List A.1 in the Appendix.

new products and processes, for applying new know-how generated by industry
and for applying research findings from the government-funded science system.
They are also the source of efficiency gains which significantly benefit sectors that
are located downstream. These especially include the service sector which has
been assuming the role of the "mother of invention" in highly developed
economies to an ever-growing degree and is evolving into a driving force behind
international innovation systems (see Section 3).

A general pattern (see also Section 2.2) can be discerned here: Germany's
technological focus is primarily on fields that require above-average amounts of
high-powered innovation effort ("advanced technology"). This focus is less
pronounced in areas that require extremely high R&D input ("cutting-edge
technology"). Today, R&D-intensive industries account for more than half of all
industrial production (50.5%[18]) in Germany. This sector's importance for the
economy as a whole is greater in Germany than in any other major industrialized
nation (Table A.2):

- Nearly 13 percent of Germany's total work force is employed in an R&D-
  intensive industry – considerably more than in the "home of the service
  sector," the USA (6%), or in Japan (9.5%), Great Britain (a good 8%), France
  (7.5%) and Italy (not quite 6%).

- R&D-intensive industries generate slightly more than 12 percent of Germany's
  gross domestic product, primarily through their strong block of advanced
  technologies (mechanical and automotive engineering, the chemical industry),
  placing Germany ahead of Japan (11.5%), the USA (8.5%) and Great Britain
  (8%).

- Cutting-edge technologies account for three percent of all people employed in
  Germany, approximately the same proportion as in Japan, and slightly greater
  than in the USA (2.5%).

The long-term trend indicates that the R&D-intensive sector's structural share is
diminishing in all major economies, albeit somewhat more slowly in Germany and
Japan than in the USA, Great Britain or France. The service sector has been the
primary beneficiary of this trend everywhere.

A growing intermeshment between different fields of technology is characteristic
of the trend seen today. For this reason, the differentiation between areas with
especially R&D-intensive production and those where production requires above-
average but not extremely large amounts of R&D does not do justice to the
phenomenon of "**generic technologies**" in certain areas (see Section **2.3**). Today,

---

[18]   This includes nearly 15.5% in the area of electrical and communications engineering,
        precision instruments/optical products, 13% in mechanical engineering/metal
        processing, 10% each in the chemical and automobile industries and 2% in office
        machines/EDP.

new, highly R&D-intensive fields – and services – evidence themselves less as "visible" final products than as "invisible" generic technologies (such as microelectronics, biotechnology, new materials, software, and microsystem technology) that have an enormous widespread impact in the form of components and intermediate products and can be traced at the goods level only with difficulty. Such fields constitute the interface between "traditional" products and new, advanced fields of application (such as pharmaceutical products/biotechnology, EDP and peripheral equipment).

It is crucial to Germany's technological performance that these "critical" basic and generic technologies be available in Germany and that the traditional strengths of German industry – advanced technologies which are frequently intensive users of cutting-edge technologies (as in the case of automotive electronics, pollution control technologies and the like) – be firmly in line with developments along technology's cutting edge.

2.2   Characteristics of Germany's industrial specialization in detail

Research and development

Although they have similar innovation capacity at their disposal, the world's major economies follow very different paths in utilizing their respective strengths. Each country sets different R&D priorities (Chart 2.1).

- Germany concentrates on maintaining a broad base and has traditionally focused its R&D efforts on advanced technology fields (automotive, mechanical and electrical engineering, the chemical industry). There has been some shift toward cutting-edge technologies (office machines/ EDP, pharmaceutical products, aerospace, measurement and control instruments) during the last decade, albeit in conjunction with a decline in overall R&D activities.

- In contrast, France, Great Britain, the USA and even Japan funnel large portions of their R&D expenditure into cutting-edge technology fields such as the aerospace industry (France, USA), telecommunications (France, Japan), EDP (USA, Japan), and pharmaceutical products (with Great Britain reporting a marked expansion of its R&D capacity).

- Furthermore, in Japan, industries that are normally less R&D-intensive, such as metal production, metal working and ship building, account for significant shares (20%) of R&D expenditure.

Chart 2.1   International comparison of R&D expenditure by sector – 1994

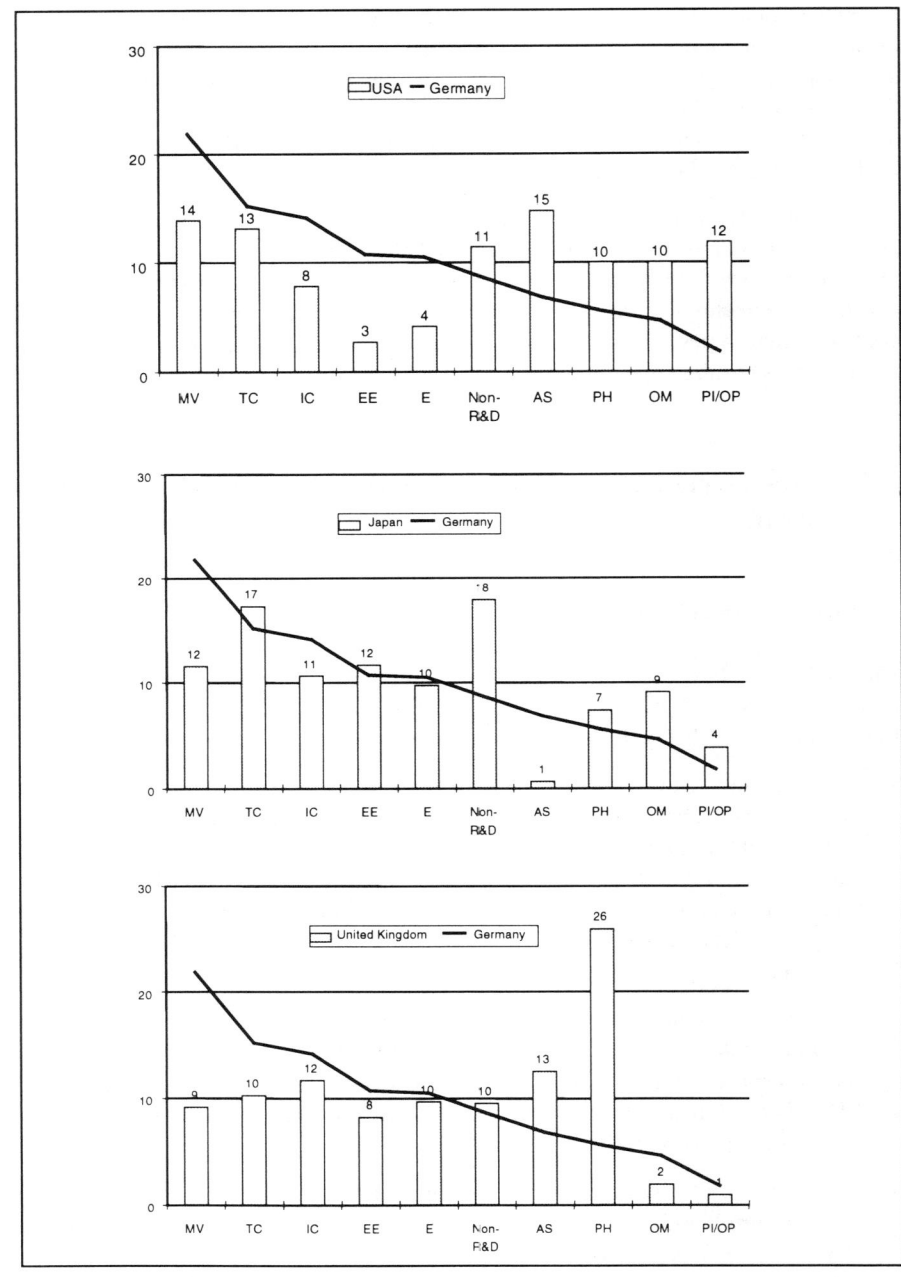

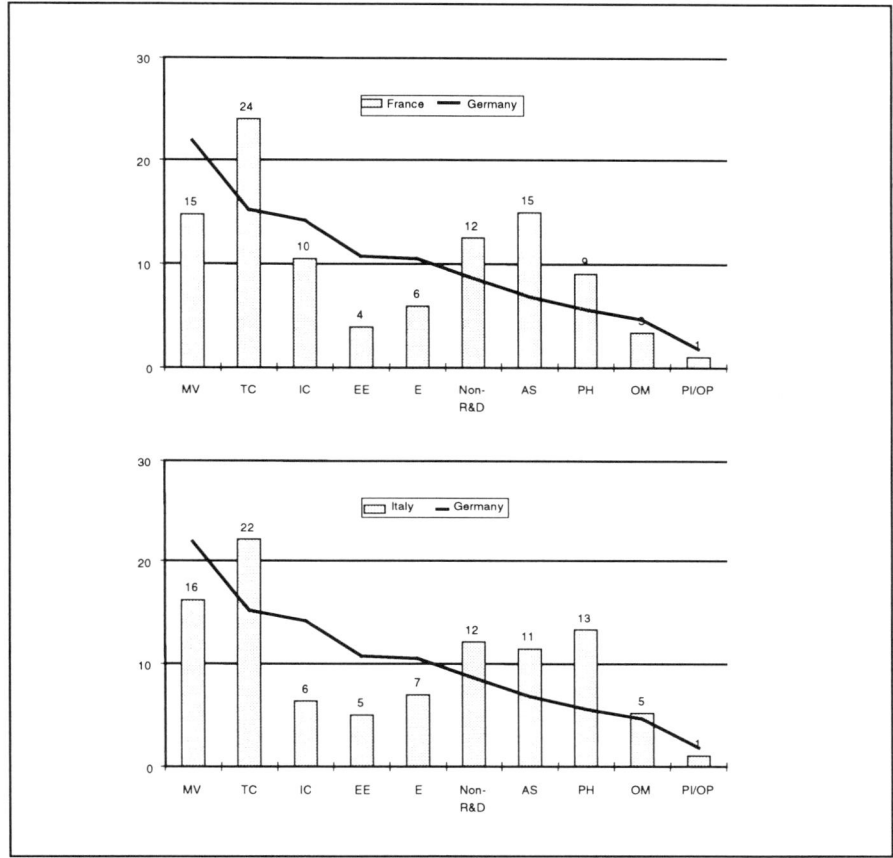

MV: Motor vehicles; TC: Telecommunications; IC: Industrial chemicals; EE: Electrical engineering (w/o telecommunications); E: Engineering; Non-R&D: Non-R&D-intensive industries; AS: Aircraft and spacecraft; PH: Pharmaceuticals; OM: Office machines; PI/OP: Precision instruments/Optical products.

Sources: OECD: STAN database, ANBERD database; DIW calculations.

- Although specialization patterns in major countries are relatively stable, the patterns found in smaller economies (Scandinavia, the Netherlands) are generally shaped by a few international corporations which set priorities selectively, primarily in cutting-edge fields of technology such as pharmaceuticals, EDP and telecommunications.

A longer-term comparison reveals that pharmaceuticals, EDP, electrical engineering and telecommunications are claiming growing shares of global R&D expenditure. Viewed in terms of industrial research, global structural change in technology is producing a shift to sectors in which, compared to other countries, Germany's R&D involvement is not as great as in other fields of industry. On the other hand, the lead that the USA, France and Great Britain have typically enjoyed in research in cutting-edge fields of technology has dwindled somewhat because governments are cutting spending, particularly in the armaments field and on large-scale projects.

Table 2.1:   Patent specialization among major industrialized countries (RPA[1])

Country	1989	1990	1991	1992	1993	1994	1995*	1996*
Cutting-edge technology								
USA	20	18	20	20	22	22	22	19
Japan	26	25	25	21	20	16	15	13
Germany**	-36	-36	-40	-36	-36	-38	-35	-31
Great Britain	-4	-3	-10	1	-2	0	0	7
France	-14	-9	-8	-12	-10	-14	-18	-18
Switzerland	-44	-45	-43	-49	-42	-35	-40	-39
Canada	-20	-2	5	6	0	21	28	27
Sweden	-11	-19	-6	-9	-8	8	9	15
Italy	-42	-44	-40	-38	-41	-36	-37	-31
Netherlands	-7	-1	0	2	-3	-10	-4	1
Advanced technology								
USA	-11	-11	-10	-9	-12	-10	-11	-11
Japan	3	5	4	8	8	7	7	8
Germany**	12	10	13	10	11	13	14	13
Great Britain	-6	-6	-9	-12	-8	-11	-14	-17
France	-4	-5	-9	-5	-4	-2	-5	-2
Switzerland	15	12	13	3	11	8	9	7
Canada	-9	-14	-3	-9	-7	-20	-27	-27
Sweden	-19	-3	-3	-9	-6	-10	-7	-13
Italy	14	18	12	10	15	11	14	11
Netherlands	-3	0	3	1	-7	6	1	2

[1] RPA (Relative Patent Activity): A positive value indicates that the share of patents in this field is larger than for total patents

* Estimates for 1995 and 1996

** Figures prior to 1991 apply only to West Germany, thereafter to Germany as a whole.

Sources: EPAT, PCTPAT; FhG-ISI calculations

Patents for the global market

Despite the recent – in part, sharp – rise in the number of patents (see Section 4.1), there has been little change in the major economies' specialization in the area of R&D-intensive categories of goods (Table 2.1). The basic positions in the international division of labor are relatively hardy and change only in small increments. Japan and the USA in particular apply for an above-average number of patents in the **R&D-intensive** field. While these two countries set the standards in this sector, Germany continues to seek a large share of patents in non-R&D-intensive fields. Germany's patent pattern has however shifted slightly toward R&D-intensive fields in recent years.

A closer examination reveals a more nuanced picture. The USA and Japan have traditionally focused on cutting-**edge technologies**. Japan however virtually fell into a "slump" a few years ago and has yet to pull out of it. Looking beyond the USA and Japan, smaller countries such as Canada and Sweden are currently reporting a positive specialization, the result of concentrating much of their R&D effort on selected, very R&D-intensive fields. Great Britain has acquired specialization advantages in cutting-edge fields of technology in much the same way. Other countries, including Germany, have yet to accomplish this. Germany has specialization disadvantages in a majority of cutting-edge technologies, particularly in major fields. However, its relative position in cutting-edge technology has seen a slight overall improvement in the past two years.

By contrast, Germany is to be found in the front ranks of the **advanced technology** field. German industry channels the greater part of its R&D resources into this area and occupies market segments for advanced-technology products. In this area, Germany holds a particularly large share of patents aimed at the application of know-how, and new scientific achievements do not play a dominant role. Instead, Germany seeks to combine established technologies and integrate outstanding achievements from science and research into traditional fields that have a very broad impact. Switzerland, Italy and, to a growing degree, Japan are also looking to advanced technology as their preferred route for maintaining and improving their technological performance.

Foreign trade

Neither Germany's patent specialization patterns nor its foreign trade specialization patterns evidence the considerable importance that the country's cutting-edge technology sector has for the **overall economy** (2.1). Rather, Germany's position in patents and on the world's markets for high-tech goods (Chart2.2) largely reflects its **industrial** structure and its focus on industrial R&D. Goods involving cutting-edge technology accounted for nearly 30 percent of all Germany's exports; goods based on advanced technology comprised the remaining 70 percent. Import statistics reveal a much more balanced picture: 42

percent of all imports came from cutting-edge fields of technology, and 58 percent can be attributed to advanced technologies.

This pattern has remained quite stable over the years: Germany is the world's largest exporter of advanced technologies with an 18-percent share of the global market (ahead of Japan with 16.5% and the USA with just under 13%). It also accounts for 11.5 percent of global trade in cutting-edge technologies (Table 2.2).

Incremental changes are only gradually taking place and tend to be evident only under close examination. The picture in the area of **cutting-edge technology goods** has become progressively favorable. The following patterns and trends are striking (Table A.3):

- Looking at the particularly R&D-intensive chemical sector, Germany is losing a large degree of its specialization advantages in the area of pharmaceutical substances.[19]

- As in years past, information technologies are still not considered to be one of Germany's particular strengths in foreign trade. However, the large share of I&C technologies in Germany's import mix does confirm the country's role as an important **user** of I&C technologies.

- The boom in the German telecommunications industry and Germany's growing attractiveness as a location for providers in the telecommunications field is largely attributable to deregulation – which Germany initiated at a comparatively late date. Germany's position on both foreign telecommunications markets and vis-à-vis import competition has improved markedly in recent years.

Although they are no longer as pronounced as during the late 1980s, Germany enjoys definite comparative advantages in **advanced technology** trade. Export and import trends ("intra-industrial trade") are moving in the same general direction. However, pronounced shifts are emerging within this sector:

- The balance of trade positions reported for chemical products, machinery, electrical engineering products, precision instruments/optical products, locomotives and rolling stock, and offices machines have tended to deteriorate.

---

[19]   This reflects the impact of the German pharmaceutical industry's innovation and growth rates during the first half of the 1990s when they were below average in comparison to the foreign competition. A number of other important positions in Germany's balance of trade – including medical diagnostical en-gineering and instruments, optical products, fibers, advanced electrical engineering, inorganic chemicals and food processing machines – are not backed by corresponding strengths in the technological field.

Chart 2.2:   Germany's technology and trade portfolio in R&D-intensive goods

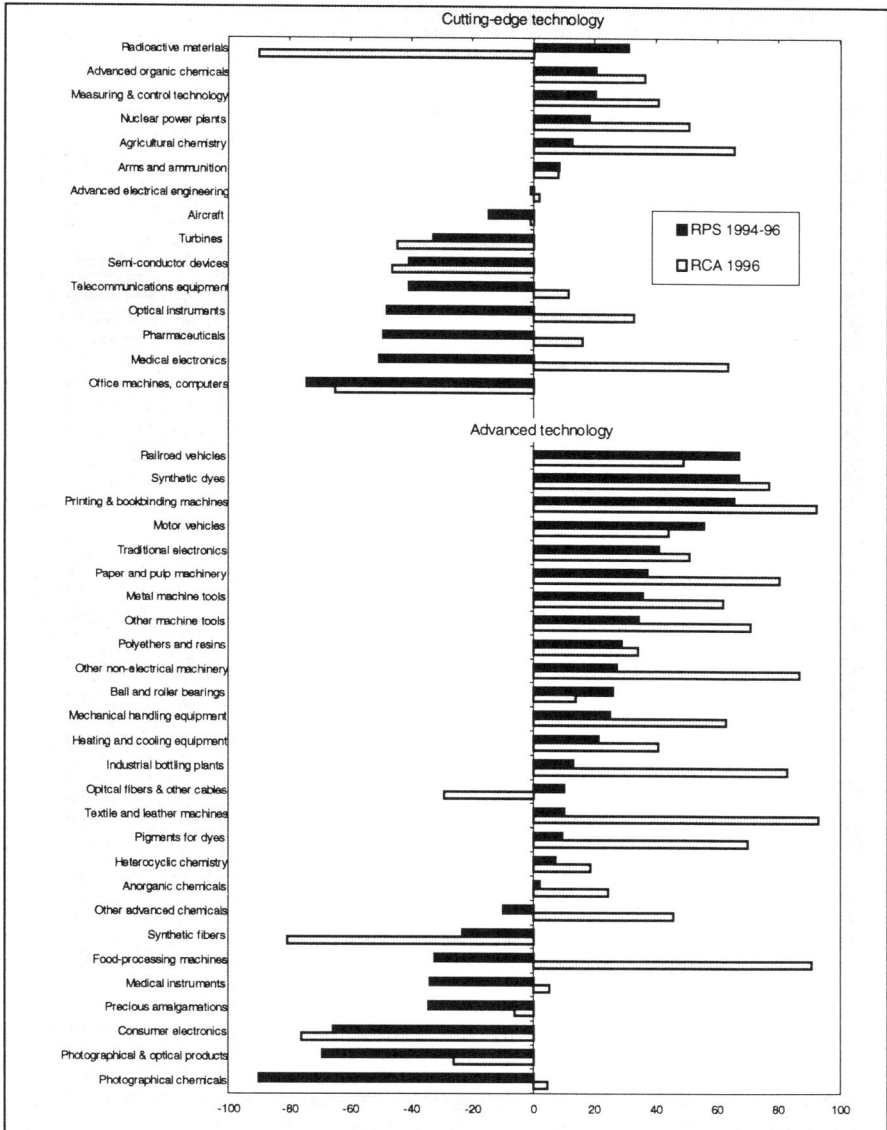

RPS (Relative Patent Share): A positive value indicates that the share of patents in this field is larger than for total patents.

RCA (Revealed Comparative Advantage): A positive value indicates that the export-import ratio is higher for this product group than for total manufactured goods.

Sources: OECD: Foreign Trade By Commodities, CD-ROM; NIW calculations; EPAT, PCTPAT; FhG-ISI calculations.

- This trend was however offset by greater export surpluses for machinery for particular industries and other non-electrical machines, optical instruments, measuring and control technology and, most particularly, for automobiles.

Table 2.2:   World trade shares in R&D-intensive goods held by OECD countries (in percentage) – 1995 and 1996

Country	R&D-intensive goods		Cutting-edge technology		Advanced technology		For information: Total manufact'd goods	
	1995	1996	1995	1996	1995	1996	1995	1996
Germany	15.9	15.4	12.2	11.6	18.3	17.9	14.8	14.2
France	7.0	7.0	8.0	8.⁻	6.4	6.3	8.0	7.8
Great Britain	7.0	7.7	8.4	9.4	6.1	6.6	6.3	7.0
Italy	4.7	5.0	3.0	3.2	5.9	6.2	6.7	7.1
Belgium/Luxemb'rg	3.6	3.6	2.2	2.3	4.5	4.4	4.7	4.6
Netherlands	3.8	3.4	4.5	3.7	3.4	3.3	4.7	4.3
Denmark	0.9	0.9	0.7	0.7	1.0	0.9	1.3	1.3
Ireland	1.4	1.7	1.6	1.8	1.3	1.6	1.2	1.3
Greece	0.1		0.1		0.1		0.3	
Spain	2.2	2.4	1.1	1.3	2.8	3.1	2.5	2.8
Portugal	0.4	0.5	0.2	0.2	0.5	0.6	0.7	0.7
Sweden	1.9	2.1	2.0	2.3	1.9	2.0	2.1	2.2
Finland	0.8	0.8	1.0	1.1	0.6	0.6	1.2	1.1
Austria	1.1	1.1	0.7	0.7	1.4	1.4	1.6	1.7
Switzerland	2.8	2.7	1.9	1.9	3.3	3.2	2.4	2.3
Norway	0.3	0.3	0.3	0.3	0.2	0.3	0.6	0.6
Iceland	0.0	0.0	0.0	0.0	0.0	0.0	0.0	0.0
Turkey	0.1	0.1	0.0	0.1	0.2	0.2	0.6	0.6
Poland	0.2	0.3	0.2	0.2	0.3	0.3	0.6	0.6
Czech Republic	0.2	0.4	0.1	0.2	0.3	0.5	0.5	0.6
Hungary	0.2	0.2	0.2	0.2	0.2	0.2	0.4	0.4
Canada	4.0	4.0	2.4	2.7	5.1	4.8	4.6	4.6
USA	16.5	17.6	23.0	24.9	12.3	12.8	14.5	15.1
Mexico	2.3	2.8	1.8	2.1	2.7	3.2	2.1	2.4
Japan	18.1	16.3	18.0	15.9	18.1	16.5	12.9	11.6
Korea (Republic of)	3.9	3.4	6.0	4.5	2.5	2.7	3.7	3.6
Australia	0.5	0.5	0.4	0.4	0.5	0.6	1.0	1.0
New Zealand	0.1	0.1	0.1	0.1	0.0	0.1	0.3	0.3

Sources: OECD: Foreign Trade By Commodities, CD-ROM; NIW calculations

Germany's specialization deficits in advanced technology goods generally center on its trade with the USA and Japan. Despite seeing foreign trade shares lose ground, **Japan** was the only industrialized nation to once again report decidedly strong specialization advantages in both advanced technologies and cutting-edge technologies in 1996. The USA traditionally has the greatest specialization advantages by far in goods requiring very high R&D intensity, due in large part to its substantial R&D activity in military-related areas. Other notable net exporters of cutting-edge technologies include not only Japan but also **Great Britain** and **France** where research in military-related fields is a comparatively large factor and is correspondingly reflected in their large world trade shares in the aerospace field.

Germany is the leading supplier of technology goods for its neighbors in Europe, even in cutting-edge technology goods (with the exception of the aircraft, aerospace, office machines/ EDP and telecommunications fields). However, Germany's strong focus on the west European market also limits its growth potential because it does not have a very strong presence in R&D-intensive-goods segments of the fast-growing American market.

Summary

Germany's specialization in "advanced technologies" is evident in all indicators: in research and development, patent volume and its foreign trade patterns. The cutting-edge technology sector, on the other hand, is not to be found on the list of Germany's primary areas of focus. Ultimately however, it is employment and earnings levels that count when evaluating a technology portfolio. Measured in these terms, Germany used to be – at least until the end of the 1980s – more successful in putting its advantages in the international division of labor in technology to use than it has been in the 1990s. This success was due to making intensive use of (often imported) cutting-edge technologies and quickly and rigorously translating them into jobs and value added (see also Section 1.2.2).

On the other hand, given that product life cycles are becoming increasingly shorter, it is not certain whether adjustment time frames will be long enough in the future for translating know-how into marketable products. Compared to a strategy which allows more flexible response to technological, occupational and sectoral structural change, a strategy that is based primarily on translating know-how and on catching-up processes bears greater risk for earnings and productivity development and for maintaining high employment levels as structural change proceeds at an ever-faster pace. Furthermore, being able to take action immediately during the initial phases of technological development also reduces the risks to growth and employment in such a situation.

2.3 Generic technologies

Only a few examples are needed to illustrate the extent to which the German innovation system can translate new generic technologies into applications, the speed at which it accomplishes this, and its outlooks for economic success with these activities.

- The process Germany follows – which is determined by the country's characteristic traits and behavior patterns – in generating new technologies and realizing their market potential can be reconstructed using the examples of biotechnology and software development.

- Similar processes can be examined in the field of microsystem technology, which is depicted here using Germany's degree of involvement in the progress achieved in this area and its share of inventions with market potential. In this field however, the uncertainties involved in translating findings into marketable products are even greater.

Common to each of these examples is the fact that "radical" changes have been accompanied by new technological challenges which Germany has taken up only hesitantly. It has answered this challenge not with "radical" innovations but rather with "cumulative" innovations that follow its well-established pattern. In other words, the catching-up process exhibits similar patterns not only within the macroeconomic aggregate but also in behavior at the microeconomic level and in mesoeconomic results.

Further, one of Germany's typical strengths – translating existing know-how into problem-solving expertise, successful products for the global market, and added value – is illustrated here using the example of environmental engineering, an "established" generic technology.

2.3.1   Biotechnology and software

Germany's characteristic innovation pattern also defines the country's advantages and disadvantages in new fields of technology and, as a result, determines corporate innovation strategies in these areas. German companies find it easier to acquire advantages in those segments where it pays to develop products and processes which follow well defined trends and offer broad applications. This applies in particular to the development and marketing of diversified, high-quality products. This pattern can be observed not only among lateral entrants (companies that are already operating successfully in other fields of technology) but also among young companies and start-ups, and is particularly evident in the **biotechnology** and **software** sectors. These two fields of technology underwent fundamental renewal during the 1980s and 1990s which opened up new technological avenues for the companies operating in them and necessitated the development of new innovation strategies.

This development in the 1980s and 1990s triggered a reappraisal of the advantages and disadvantages of internal product development as opposed to conducting product development within a collaborative framework or external networks with other high-tech companies. Pharmaceutical firms have traditionally conducted basic research and product development in their own labs with their own research personnel. In much the same way, German companies have also depended on their own EDP departments to develop software. Technological change led to the outsourcing of R&D activities which the respective company had formerly conducted itself: Small biotech firms now offer established pharmaceutical corporations not only promising biotech products but new research impetus as well, while external companies that produce standard software in conjunction with IT-service providers cover an ever-growing share of the software demand from German companies.

Revolutionary new genetic engineering processes for developing medicines have given rise to scientific and organizational challenges. New biotechnological research methods have necessitated collaboration with smaller, more dynamic firms. In the software field, the emergence of personal computers, client/service architectures and open operating systems has led to a situation in which ever-larger portions of internally developed dedicated software are being replaced by highly sophisticated standard software products. Today, external software service providers are often responsible for installing and servicing these standard products. German companies have reduced the share of their internally-developed, dedicated software in the information technology (IT) field and draw a growing portion of their software solutions from external product markets and external IT service providers.

The "German innovation system's" response in both cases was **initially very hesitant**. The German biotechnology field lagged substantially behind its British and American counterparts during the 1980s. Much of the German computer industry was virtually overwhelmed by the magnitude and speed of the change seen in both technology and the market. However, now would appear to be the time to **reassess** the German innovation system's strengths and weaknesses in the software and biotechnology fields. In the field of biotechnology, active federal and state funding programs and the amendment of the Genetic Engineering Act have enhanced Germany's attractiveness as a location for business and industry; a large number of new biotech companies have been and are presently being founded. Endogenous adjustments in the software industry – which were conducted without government assistance – also demonstrate that Germany certainly can be made more interesting for high-tech industries: The emergence of major German software service providers and signs of strength in the standard software field give cause for some optimism.

This development can be elucidated by differentiating between market segments on the basis of their technological features. For example, therapeutic agents and "off-the-shelf" standard software products are dependent in some cases on **discrete technological advances**. In other words, they are dependent on technological developments which have limited, highly specialized applicability and very short-lived market prospects. The value of these discrete technological developments diminishes rapidly because they are quickly overtaken by new, alternative products and processes. In contrast to this, platform technologies in the biotechnology field and in the software service area tend to be characterized by **cumulative technological product and process developments** whose value remains stable over a longer period.

To what extent are conditions in Germany suited to these new industries and how have firms responded to this development in their innovation strategies?

- **Firstly**, major German corporations began by reorganizing their R&D activities to draw greater benefits from external alliances. As an initial step, they established international alliances with American and other foreign technology-based companies. On no account should such alliances be considered proof that Germany has deficits as a location for business and industry: Studies on German multinational firms show that German industry is quite capable of profiting from international R&D specialization by making use of cross-border cooperation.[20]

- **Secondly**, alliances are now being established between German companies as well because small, new R&D-intensive companies are also being founded in this sector in Germany. New market structures emerged in these high-tech industries over the years and German companies were able to position themselves within these structures. The chances for German start-ups to get a foot in the door to these markets also increase as these new industries become older. For example, many successful German companies have specialized primarily in a number of broad platform technologies whereas US biotech firms have concentrated not only on platform technologies but also on the production of new medicines to treat illness (therapeutics). Although the media generally emphasize those software companies that mass-produce software (which is more typical of the US economy), there is an enormous market for business software where German software companies have been able to establish themselves in the area of highly sophisticated standard software and in software service. Both these cases show that the special advantages of German companies are to be found in the area of diversified high-quality products with potential for cumulative improvement.

Different technological traits require different kinds of expertise in order for innovation to be successful:

- On the one hand, companies working in the area of therapeutics or standard software need enormous amounts of venture **capital** due to the uncertainty of the market and the sector's high failure rates. By contrast, the financial risks involved in platform technologies and software services are not as sizable.

- On the other hand, the fact that technology in these industries develops cumulatively necessitates a willingness on the part of companies and their employees to invest in company-specific **capabilities** whose advantages will become evident only after some time.

---

[20]   Cf. J. Cantwell and R. Harding (1998), The Internationalisation of German Companies' R&D. See also work regarding the globalization of industrial R&D in selected fields of technology which was conducted as part of the 1997 report on Germany's Technological Performance by FhG-ISI/DIW/ZEW (1998).

The following table (Table 2.3) outlines the differences arising from these technological traits.

Table 2.3:   Market segments in the biotechnology and software industries

	Therapeutic agents, standard software products	Platform technologies, software services
Direction	Development of new products to satisfy the needs of the mass market	Creating expertise with broad applicability
Type of technologies	Discrete technologies	Cumulative technologies
Level of company-specific know-how	Low	High
Level of financial risk	High (technological and market risks, high R&D costs)	Low (clearly defined markets and technologies, lower R&D costs)
Level of general market risk	High	Low

Germany's involvement in both biotech platform technologies and software services can be explained by its traditional institutional framework which promotes not only incremental innovation, but long-term relations between companies and owners, and the accumulation of know-how and experience as well. By contrast, the therapeutical agent branch of the biotechnology sector and the standard product area in the software field are characterized by more radical innovation, shorter time horizons on the markets, and limited time frames for making use of innovative know-how.

This does not however mean that German companies cannot hold their own against foreign competitors in these areas. Germany most certainly has a number of successful pharmaceutical manufacturers as well as successful software companies and promising start-ups in the standard software market segment. The point here is only that it takes more effort to establish German companies in riskier, discrete technologies in the high-tech field, whereas lower risk, cumulative technological developments constitute a more consistent fit with the German innovation system and are therefore easier to realize within the institutional framework Germany offers for innovation activity. Germany's strength lies in translating "radical" innovation into advanced products and processes, following the lead of the "lead user."

2.3.2   Microsystem technology

Viewed as a key technology for the coming century, microsystem technology[21] is expected to offer a multitude of market opportunities and achieve broad diffusion in many areas of technology. In light of this, the Federal Ministry of Education and Research has been fostering microsystem technology since 1990 by means of special programs.

Only in the years since the late 1980s has the worldwide patent volume in this field reached an appreciable level, with growth remaining moderate during the 1990s. The number of scientific publications in this field also began to pick up toward the end of the 1980s. Their number is however growing markedly faster than the number of patents. Accordingly, microsystem technology is still in its early stages, with a focus on basic and applied research that is geared to the long term. At international level, Germany ranks second behind the USA in its number of publications and patents. Federal assistance that was launched in 1990 has contributed in particular to helping Germany's research efforts in this field keep pace with international developments: The number of publications from Germany has increased sixfold in the last five years. At 300 percent, the increase in the number of application-specific patents was more moderate (Chart 2.3). Based on this, exploitation opportunities are more modest.

Since microsystem technology is a generic technology rather than a product, precise production and trade figures do not exist for it. Market analyses currently assume world market volume to be approximately US$ 2 billion and expect this to range between US$ 4 billion and US$ 14 billion for the year 2000, making any prognosis very uncertain. In any event, market volume in this sector is still moderate compared to other key areas such as microelectronics. Up to now, the microsystem technology market has been buoyed by just a few products such as read-write heads for disk storage, heads for ink-jet printers, acceleration sensors for airbags and sensors for measuring blood pressure. Experts expect new revenue-generating market segments to open up for peripheral EDP equipment, biomedical microreactors and optical communications (optical switching, optical "connectors"). An analysis of patents indicates that German firms are up to international standards in their degree of application-orientedness in the area of microsystem technology; deficits can however be observed in the optics fields, particularly in comparison to the USA. This could have a negative impact on Germany's future market prospects.

---

21   The term "microsystem" is used to designate a module whose functional elements exhibit microstructures and in which several of these elements work together on a systematic basis. Electrical and mechanical functions are often coupled in this process which is why technical literature also speaks of micro-electromechanical systems.

Chart 2.3:   Comparison of EPA patents and SCI publications of German provenance on microsystem technology

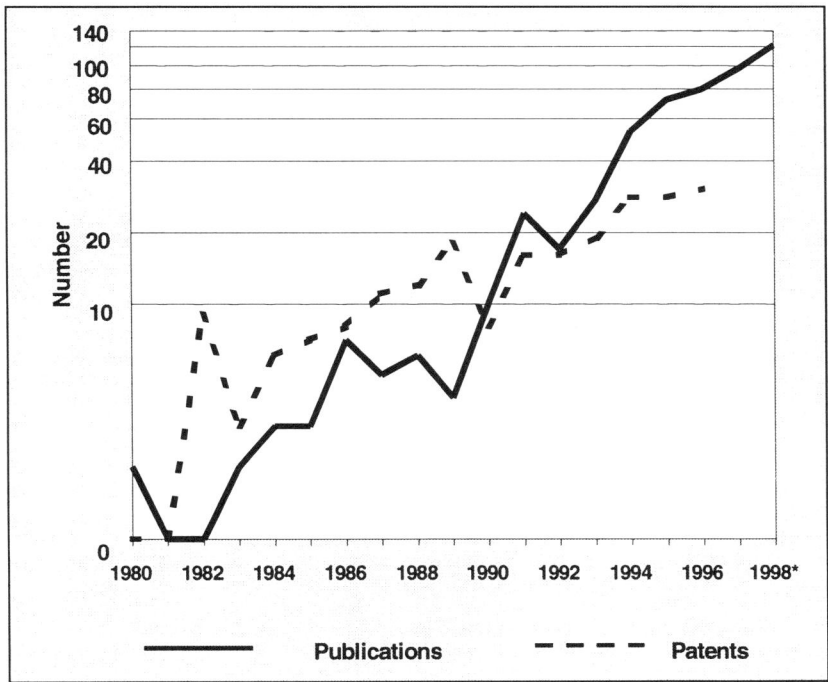

*   Estimated

Sources: SCI; FhG-ISI calculations

Germany has currently acquired a good starting position for itself in microsystem technology, both as a science and as a technology. In coming years, it will be essential to Germany's technological performance that more be done than in the past to create possible applications for microsystem technology in Germany's traditionally strong fields. Germany is not however a "lead customer" in the EDP periphery applications that have been dominant to date. Although Germany has high levels of expertise to offer, it needs markets where it can put it to use. One gets the impression that although the funding in this field is highly application-oriented, very few firms have taken up this new technology. As a result, research institutes that are likely to have long-term application goals have become major players in the microsystem technology field. Without wanting to question technology policy's concentration on important non-industrial research require-ments in the medical sector, it should be noted that funding would produce tangible success more easily if it were to assign greater attention to **markets** that grow on a short and medium-term basis and offer German companies opportunities as well. Application-oriented funding should also provide proof of the economic benefits it brings.

2.3.3   Environmental technology

New technological developments are often absolutely essential to solving environmental problems. However, solutions that are based on environmental technology tend to require the optimal application of existing know-how to specific environmental problems – which happens to be Germany's particular strength – rather than the development of cutting-edge technologies.

The rather sharp drop in worldwide patent applications in the **environmental technology** field was probably due primarily to the phasing out of government rehabilitation measures (additive environmental protection) in the 1990s, the lack of adequate incentives to pursue ongoing (precautionary or integrated) environmental improvement and a corresponding decline in private interest (due to falling prices for raw materials, for example).[22] Although Germany's share of total patents has declined considerably, it is still decidedly large – except in measurement technology.[23] Germany's already high level of environmental protection is occasionally cited as a reason for the country's waning invention activity. The positive impact that environmental regulations – which have traditionally been geared to the "state of technology" – have had on innovation is also gradually diminishing.

Germany shines in environmental protection with a decidedly advanced range of products. Technological capability definitely pays off in this field: Germany generally imports "off-the-rack" environmental technology and exports high-quality, individually manufactured products. However, it will not be easy for Germany to continue bringing its technological advantages to bear on the **pollution control market**. Today's markets are highly segmented as a result of national legislation, a resultant domestic orientation, and the large share of investment and procurement that government accounts for. Germany exported potential pollution control goods[24] worth some DM 37 billion[25] in 1996. This

---

22   Cf. last year's report.

23   Germany still accounts for half of Europe's total patent output. Cf. last year's report.

24   The list of **potential** pollution control products which was used for these calculations and estimates was developed on the basis of the "official" list issued by the Federal Statistical Office. With regard to this method, it must be noted that only traditional "end-of-the-pipe" environmental protection technologies could be taken into account in many cases. Conclusions regarding competitive rankings in the field of modern "integrated" environmental protection – which is not yet very widespread – cannot be drawn from this information. Estimates for the year 1994 indicate that 35% to 40% of all potential pollution control goods are actually used for environmental protection ("dual use"). In this case, export volume for pollution control products would approximate the volume reported for pulp, paper and paperboard products, refined petroleum products, precision instruments/optical products/watches and clocks, plastic products and fabricated metal products.

25   This is almost 5% of Germany's total export volume for manufactured goods.

represents a world trade share of 17.5 percent which makes Germany the second largest exporter in the world (Chart 2.4), on the heels of the USA (18%). Japan follows at an unusually large distance to take third place in the list of the world's largest providers of pollution control technologies with a world trade share of nearly 13 percent.

Chart 2.4:   World Trade shares of leading countries in pollution control products 1989–1996*

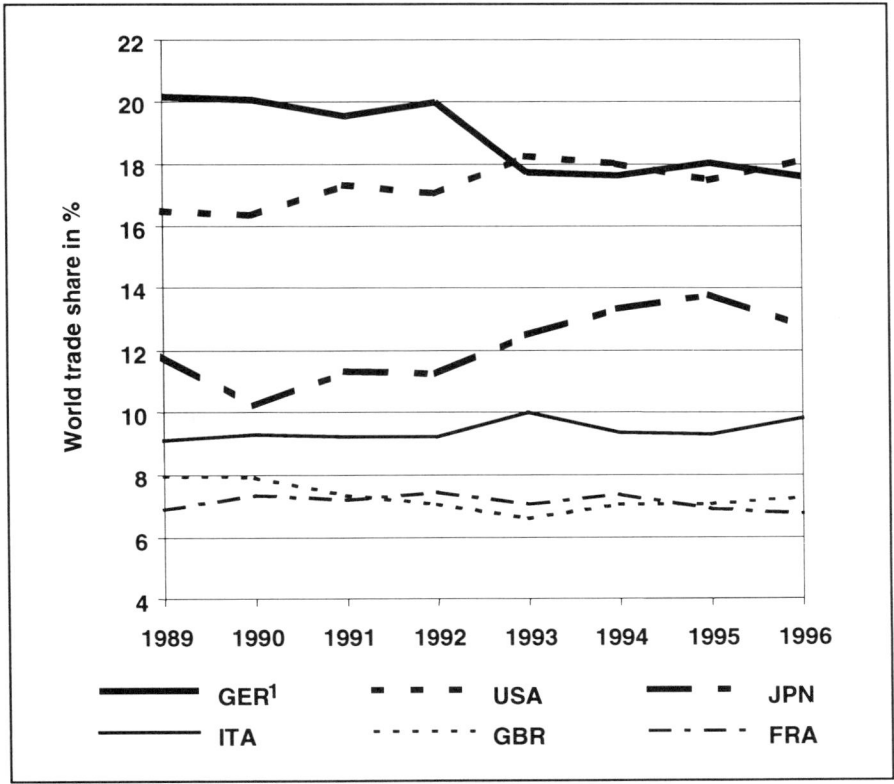

*   Data for 1989 through 1994 were calculated on the basis of the OECD-29

1   Figures are for Germany as a whole starting 1991 and are therefore only partially comparable to previous years

Sources: OECD: Foreign Trade By Commodities; unpublished data for 1989–1995; 1996 CD-ROM; NIW calculations

The importance of the individual **pollution control fields** (air quality control, water/waste water, waste management and the generic field of measuring and control engineering) in Germany's bundle of export goods closely reflects the global distribution of demand for products from various environmental media, with Germany accounting for some 17 to 18 percent of global exports in the individual areas. Despite this apparently favorable balance, it must be pointed out that full advantage is obviously not being taken of existing opportunities. With the exception of measuring and control technology, Germany's ranking in patents with world market potential indicates that opportunities for expansion and specialization still exist.

The fact that new approaches to environmental protection are not high on policy makers' agendas has clearly left its mark on the pollution control goods and services market. By shaping the regulatory framework, the state exerts decisive influence on the quality of demand for technology. As incentives to innovate wane, imitators start gaining the upper hand. However, rather than playing midfield in those areas where it can bring its special expertise to bear – namely, combining the latest scientific and technological developments with well-established strengths – Germany should be striving for top ranking and technological leadership. Otherwise, territory will be lost on "secure" markets. This would however entail establishing incentives aimed at environmental protection that goes beyond the "state of technology." Firms in Germany must be given the opportunity to prove their capabilities and to open up and develop markets. Germany will not be able to expand on international markets if it simply gears itself to the state of technology which, by definition, is not very innovation-oriented. At the same time, it should be noted that the use of environmentally friendly production processes generally involves additional costs – which could have a negative impact on the user's (international) competitive strength. For this reason, only the nuanced use of instruments to stimulate the market for environmental technologies appears to be advisable.

Despite the fact that, for a while, Germany tended to pay little attention to environmental protection and that a new surge in innovation is currently needed, all known forecasts point to **expansive market development** – particularly at international level. Stepped-up climate protection efforts – as expressed in, for instance, conventions on climate change and in individual countries' voluntary commitment to reducing their CO_2 emissions – play a particularly important role in this context. The environmental protection industry should seek solutions that can be drawn upon and applied worldwide, with an eye to fostering sustainable global growth. Furthermore, new paradigms will be replacing the old: "Visible" (additive) environmental protection will diminish in importance while "integrated technologies" which avoid generating pollution in the first place will become established. This means that old methods will be replaced by modern technology and services. Integrated environmental technologies generally offer greater innovation potential: This will favor know-how-intensive industries and advanced research, planning and consultation services.[26]

---

[26]   Cf. B. Gehrke, H. Legler and U. Schasse (1998), Regionalökonomische Effekte von Klimaschutz-maßnahmen in der Region Hannover.

3 Innovation in the Service Sector and Competitive Strength

Apart from the incentive provided by scientific progress, the incentive for innovation essentially comes from customers and suppliers. In other words, incentive is mediated through markets, their momentum and their intermeshment – in short, through the economic structure. The greatest difference in the German system when compared to the USA and other highly developed countries such as Great Britain is to be found in the service sector which is considerably less developed in Germany (see Section 1.2.2). This also has considerable consequences for the development paths and types of specialization which German industry has chosen. The German service sector is catching up only gradually. In particular, specialized business services and activities related to the banking industry and insurance business have developed greater growth momentum – albeit along what would be considered a comparatively low baseline by international standards. However, growth has been modest even in these fields in recent years.[27]

3.1 Interaction between industry and the service sector

The type of specialization that an economy has "chosen" has a decisive impact on the innovation goals and behavior patterns of the players involved in the innovation process. As a consequence of the options which the respective country's institutional framework (regulations, organizations, division of labor, legal framework, system of public finance, system of government, resources, etc.) and societal traditions offer, each individual economy has taken its own course in terms of technology and industrial structure. The innovative **pressure** exerted by intermeshed industries and the innovation **slip-stream** that follows in the wake of demand determine the direction and intensity of innovation activity. Strong demand for high technological standards has an effect on innovation activities such as R&D and on the use of new technologies and production processes.

Industry and the service sector are growing increasingly closer together as a result of their intermeshment in the marketplace (as in the case of multimedia). The "interaction theory"[28] views the service sector primarily as a customer of and provider to industry in this process:

---

[27]   Cf. also H.-H. Härtel and R. Jungnickel (1998), Strukturprobleme einer reifen Volkswirtschaft. Analyse des sektoralen Strukturwandels in Deutschland.

[28]   Cf. also H. Klodt, R. Maurer and A. Schimmelpfennig (1997), Tertiarisierung der deutschen Wirtschaft.

- Service providers are important **users** of technologies originating in the industrial sector. Impetus for innovation is communicated from the service sector to the industrial sector. Examples of this process would include the close links between medical technology or pharmaceutical engineering and the health care sector, between telecommunications and telephone corporations, and between aeronautical engineering and airlines.

- Business services play a particularly important role in technological performance. They reinforce the **link** between industry and the technology sector, and simultaneously enhance industry's competitive strength by supplying know-how. As a rule, business services expand fastest in areas where demand their products exists among innovative industries.

Performance in the industrial sector is dependent upon the service sector and vice versa. German industry's strong position on international markets would be virtually inconceivable without the existence of high-powered services at its doorstep. New "networks" are emerging in growing numbers between innovative industrial companies and service providers. Consequently, the type of specialization a country has chosen is of extraordinary importance for the innovation goals that the industrial and service sectors set themselves and for the impetus they have to pursue innovation.[29]

- In **non-R&D-intensive industries** and in fields of **advanced technology**, innovation serves first and foremost to protect and expand the share of what is usually a not very dynamic market. Advanced technology follows well defined trends which hold strong promise of generating a return on the R&D investments involved: The quality of existing products is improved, production processes optimized and labor, material and energy costs are lowered through rationalization. Innovation often takes the form of product differentiation via technology; the portion of "cumulative innovation" is generally quite large. Germany is decidedly strong at international level in this area and is able to secure significant leads over its international competition. Advanced technology companies draw their impetus for innovative activity primarily from other firms working in the advanced technology field and from industries that are less R&D-intensive.

- By comparison, radical innovations usually account for a large portion of the innovations generated by **cutting-edge technology** which typically develops new markets and discovers new areas of application. Labor costs play a subordinate role in this field. Cutting-edge technology is geared more strongly to dynamic growth. Launching new technological trends and establishing fundamentally new products on the market are the decisive challenges facing such firms. Not only are the R&D costs enormous, but the chance of market

---

[29]   Cf. regarding the following observations G. Licht and H. Stahl (1997), Ergebnisse der Innovationserhebung 1996.

success is also extremely uncertain. Companies working in the cutting-edge field are highly dependent upon the willingness of the service sector and of the state to adopt innovation.

3.2   Innovative services and competitive strength

Complementarities both between the service sector and cutting-edge technology and within the advanced technology field evidence various strengths and weaknesses in the German innovation system: The advanced technology field profits from German industry's large share of and strong demand for innovative capital and intermediate goods. Parallel to this, the application-oriented part of government-funded research strongly supports advanced technology (as manifested by the large portion that engineering sciences represent at universities and *Fachhochschulen* – practice-oriented technical colleges). By contrast, it would appear that Germany's service sector (which by international comparison seems to be somewhat less well-developed and less innovative than in other countries) does not "demand" as much of the innovative area in the cutting-edge technology field.[30] Which is also why Germany does not specialize at international level in cutting-edge technology. However, the importance of innovation is growing in tandem with the share the service sector represents in the economy.

Services in the international arena

The **percentage of companies** in the service sector that **generate innovation** is approximately as high as in the industrial sector. The service sector is also becoming increasingly important for the country's overall innovation potential. A number of indicators (start-ups, training, innovation activity, etc.) show that services are becoming more actively and continuously involved in innovation efforts and are shaping them to an ever larger degree.

The service sector's innovation activities not only impact industry's innovation capability ("interaction theory"), they have also developed into a vital pillar for the international competitiveness of the service sector itself. In many cases, services no longer have to be rendered at the recipient's respective location. New I&C technologies in particular make it possible to do business in services (such as in the software, airline reservation systems, telebanking, telemedicine and insurance fields). Although service exports – which are usually coupled with goods (such as transport services) – are not growing at a faster overall pace than goods exports, primary services are increasingly the object of **international trade**

---

[30]   Cf. also M. Grömling, K. Lichtblau and A. Weber (1998), Industrie und Dienstleistungen im Zeitalter der Globalisierung.

(triggered, *inter alia*, by the liberalization and privatization of government services such as telecommunications, the railway and the energy supply) and report extremely high growth rates. Globalization and worldwide efforts toward deregulation (such as GATT) are independent forces that also drive the service sector's growing innovation-orientedness. With services becoming increasingly saleable, the importance of securing proprietary rights and rights of exploitation – **copyright protection** – is also growing because patents are often not suitable for protecting service innovations.

As service markets become internationalized, the resultant increase in competitive pressure leads to technical progress that enhances productivity. Innovation – regardless of whether it is product, process or organizational innovation – is of increasing importance for export performance in the service sector.[31] The more innovative service providers are, the greater service export volume will be. The technology-intensive and know-how-intensive service sector is on its way to becoming as important for technological performance as (R&D-intensive) industry is. There is however an unmistakable trend toward standardizing services. This trend is leading to the increasing use of "industrial" production methods in the service sector as well and enables a further increase in the international division of labor. The service sector is also moving "simple" activities (in software production, for instance) to other locations.

Today, some 20 percent of all service providers in Germany export their products. However, more than 30 percent of Germany's service providers already have international competition, a situation which tallies with Germany's large trade deficit in services. In terms of international trade in goods and services, Germany has **comparative disadvantages** in service trade. This is well illustrated by Germany's trade with the USA[32] which has particularly grown in the area of insurance services (trade volume in 1997: US$ 1.5 billion) and technical services (US$ 1.3 billion). Germany enjoys specialization advantages vis-à-vis the USA only in the area of insurance services. It reports comparative disadvantages and even import surpluses in all other areas.

---

[31]   G. Ebling and N. Janz (1998), Export Behavior and Innovation Activities in the Service Sector - Empirical Results for a Cross-Section of German Firms.

[32]   Available international trade statistics are not suited to analyzing service trade broken down by individual services. Technology-oriented service trade is conducted largely within multinational corporations which set their own internal costing prices. Since multinationals often operate production locations in a variety of countries, statistics on such trade do not accurately portray the comparative national advantages and disadvantages involved in service production. Only the trade statistics from the USA provide precise information regarding the service trade between non-associated companies.

Use of technology and impediments to innovation

The service sector's innovation capability is primarily – but not exclusively – shaped by its use of I&C technologies. Based on their use of technology, service companies can be divided into two different **innovation categories** (Chart 3.1 and Chart 3.2):

- **I&C-intensive services** (such as software companies, banks, insurance companies) deploy only I&C technology, albeit very intensively: I&C technologies are used primarily to improve internal processes and enhance service quality. They are however also used to help in developing new business segments. I&C technologies are of key importance for R&D particularly when they are used in areas of core competences or constitute the basis for the business itself. For I&C-intensive companies, the incentive to be innovative comes from customers in the service sector.

- **Multitechnology-intensive services** (such as telephone companies, railways) use technologies such as traffic technology, biotechnology and environmental technology in addition to I&C technologies (to save labor, cut energy costs or reduce pollution, for instance) or even **develop** these technologies themselves, often citing industry customers as the source of their impetus for innovation.

This differentiation is significant insofar as these two groups differ substantially with respect to where and how they obtain stimulus for innovation, how they collaborate within the innovation process and with respect to the impediments they must deal with (see in this connection Chart 3.4 and Chart 3.4).

Germany has to strengthen its innovation capability in the service sector if it is to maintain its level of technological performance – in both innovation-intensive industrial goods and in services – in the future as well. However, a substantial portion of Germany's innovation potential lies fallow because services are more highly regulated here than in other countries, with the general effect of putting a damper on innovation activity. Foreign investors in particular bemoan Germany's thick regulatory maze and the requirements it imposes on market access in precisely those parts of the German service sector that are geared more to "public needs" (such as health care, power supply, transport and public utilities).

The percentage of firms whose innovation projects are hindered by impediments is usually higher in the service sector than in industry (Chart 3.1). **Financing** in particular is one of the largest problems in the service sector where an especially large number of small firms are innovative and lack material security. Banks and EDP service providers are particularly hard hit by the **shortage of computer scientists** which has been observed in recent years.

Chart 3.1:   Impact of impediments on time frames for innovation projects in the service sector: relative importance

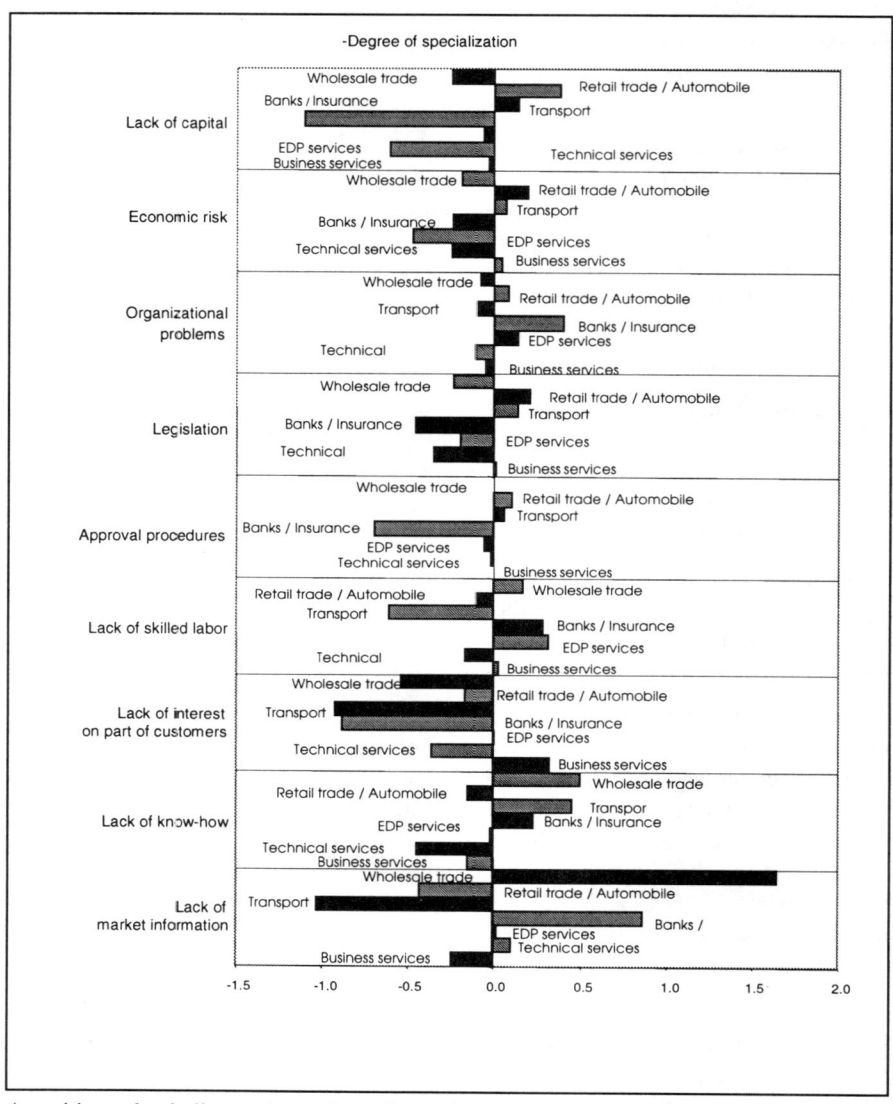

A positive value indicates that an impediment is above-average importance

Source: ZEW/FhG-ISI: Mannheimer Innovationspanel.

Commerce has more than an average amount of problems with a lack of **technical information**. At the same time, this sector is an important buyer of banking services. The banking sector however belongs to that group of service providers who bemoan their **customers' lack of interest** in innovation more often than other service providers do. This would indicate a rather conservative demand for banking services on the one hand, and the provision of meager technical support on the part of the scientific community and outside suppliers on the other. The lack of technical know-how and market information leads to an above-average number of delays in innovation projects in this sector.

Chart 3.2:   Goals pursued by innovative service providers, by type of technology used

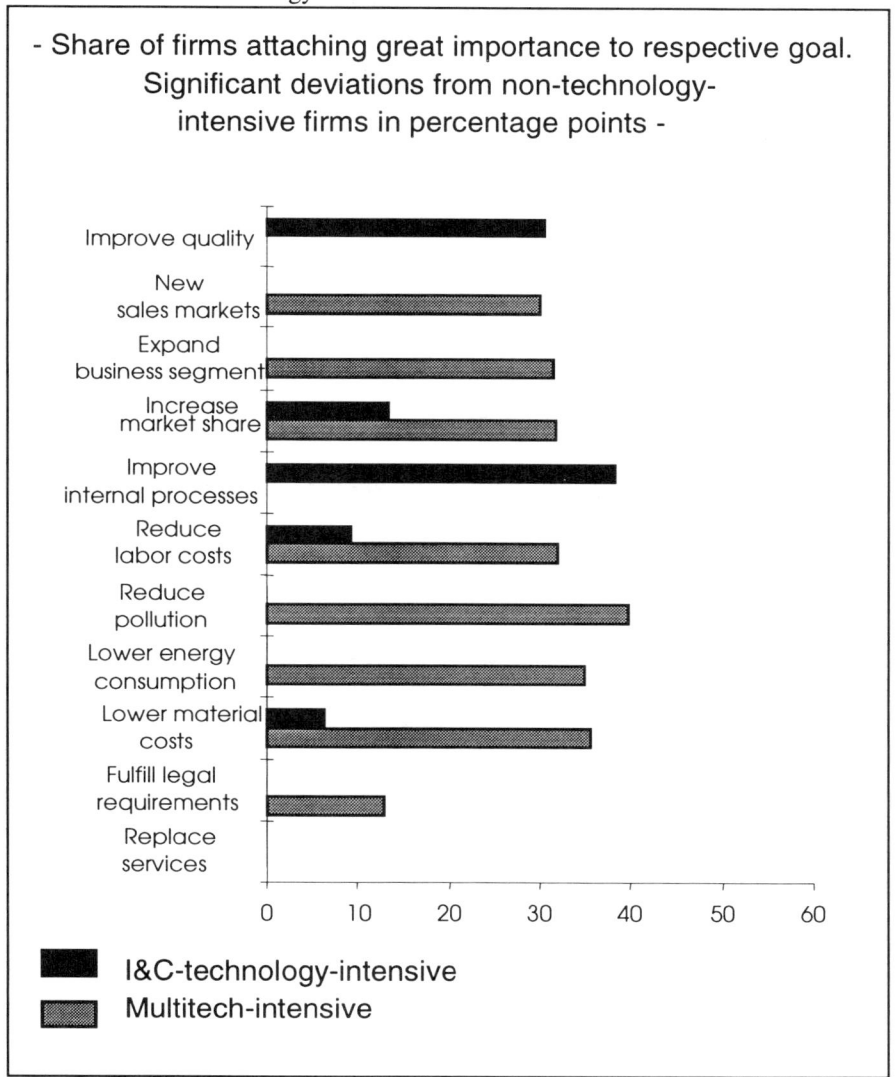

Source: ZEW/FhG-ISI: Mannheimer Innovationspanel

Chart 3.3:   Sources of information for innovative service providers, by type of technology

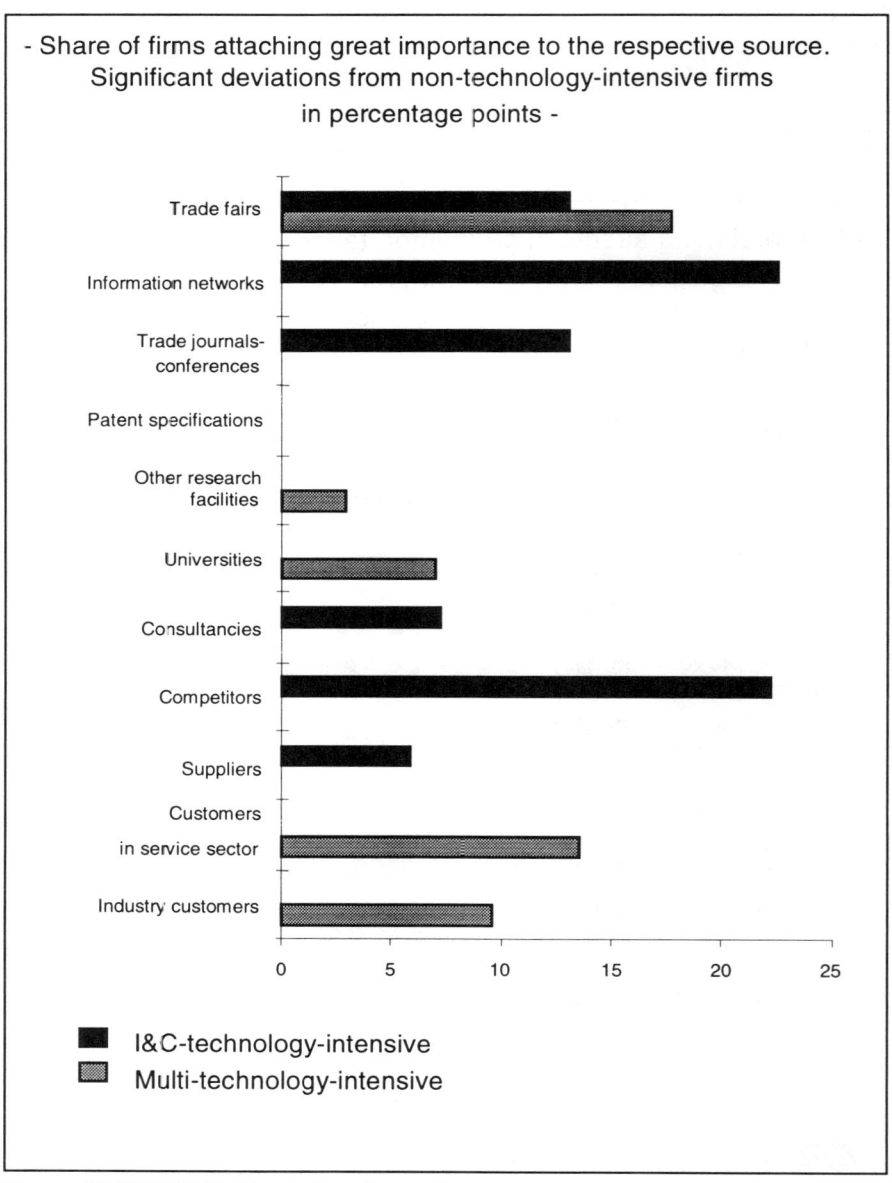

- Share of firms attaching great importance to the respective source. Significant deviations from non-technology-intensive firms in percentage points -

Source: ZEW/FhG-ISI: Mannheimer Innovationspanel

Within the value-added chain, "contributor" shares are shifting from the hardware to the software field. Which means that services are dependent on the application of advanced technical solutions. In this regard, it is essential that more attention be paid to the conditions governing the adaptation of technical know-how in the expanding service sector (research assistance, R&D infrastructure, and application-related assistance and funding) because in reality only those firms that make intensive use of "multiple technologies" find broad **support** for innovation in the scientific field – in other words, from universities and other research institutions.

Put in other words, this means that stimulus coming from the marketplace is the primary factor regulating the service sector's innovation activity: Customers from the service sector are the foremost source of demand for service innovation. Therefore, the service sector's inherent dynamism spawns innovative impetus which radiates to industry's cutting-edge technology fields. Service companies respond to these interdependencies by increasing their involvement in innovation projects: Collaboration is growing within the service sector while collaboration between the service sector and industrial companies is on the decline (Chart 3.3 and Chart 3.4).

Special aspects of the R&D process in service firms

In principle, R&D is of relevance to services not only because it ensures the development potential of the service sector itself, but also because it integrates that part of the service sector that conducts research into the national innovation system (committed research, teaching and instruction, etc.). However, as a force behind innovation, R&D activities play a much smaller role in the service sector than in the industrial sector. Research and development are conducted almost solely in the software and communications fields (telecommunications, for example). And as a rule, even large service companies do not have R&D departments or permanent R&D staff to conduct their research and development activities. Instead, project teams of varying composition conduct R&D activities which are often part of daily routines and closely linked to specific projects.

The nature of R&D conducted in the service sector differs in a number of ways from that of R&D conducted in the industrial sector. In essence, service R&D is less technology-oriented than R&D conducted in other sectors. And even when a new product is the object of research, as in the development of a new banking service, the product itself is seldom a technical innovation. Rather, new service products are developed using technically modified resources from other sectors. Other important research subjects in the service sector involve organizational concepts, social competences, market acceptance research and pilot projects. Although R&D projects aimed at developing new organizational concepts increase our knowledge in the economics and sociology fields, they do not contribute to an increase in our knowledge of technologies.[33] Such projects also generate important know-how about potential applications for new technologies.

Chart 3.4:   Share of firms collaborating on innovation projects 1996

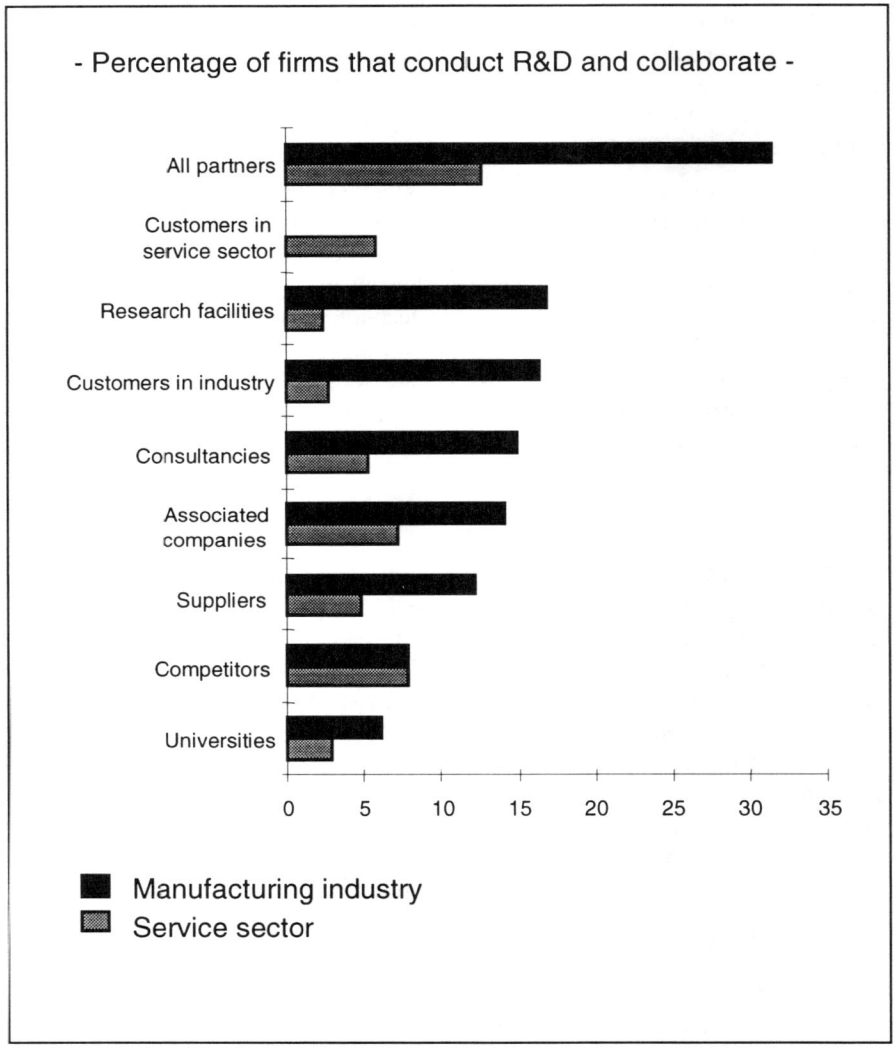

Source: ZEW/FhG-ISI: Mannheimer Innovationspanel

---

[33]   Extensive research on best-practice methods in the service sector provided the basis for models that other countries have used successfully to improve service quality and service management.

Improving the methods used in market research is another important research goal for service providers, whereby it must be remembered that it is difficult to demarcate R&D from other innovation activities. For this reason, R&D staffing levels and R&D statistics from the service sector do not match those from the industrial sector in their ability to act as an indicator for innovation, and are even less suited for use in international comparisons.[34] The percentage of highly qualified employees out of all employees is more likely to provide a relatively good idea of a company's innovation capability.

3.3   Summary

The service sector is developing increasingly into a **pivotal factor** for an economy's innovative strength. The service sector's **inherent dynamism** generates an impetus that calls for top-level technological achievements and places strong demands on industry's innovation capability. The German service sector is however lacking in inherent dynamism; new, sophisticated and high-powered markets are not being developed in the service field – even though the pool of services available to the population is relatively small. Without momentum, there are no challenges to the innovative ability or the technological capability of the German innovation system. The lead market function normally assumed by the service sector often lies dormant in Germany.

Education, science, research and technology policies alone are not enough to coax out innovation – even when they are pursued with rigor and perseverance. Attractive market potential for goods and services is a decisive prerequisite for the development of advanced innovations in Germany.

It is no accident that the signals coming from Germany's service sector are not as clear as those found in other highly developed economies or that the German service sector's international competitive strength leaves much to be desired. On the one hand, an analysis of Germany's innovation system reveals a tendency to pursue "safe product innovations" (such as banks and insurance) rather than chance higher risk. On the other hand, many types of services in Germany have long histories as regulated, protected fields with little competition. These institutional and legal conditions are part of the country's "national innovation system." They have to be "re-regulated" and barriers to innovation have to be dismantled. Given that the service sector is frequently subject to institutional rigidity, these factors impact it more than they do industry.

---

[34]   It is likely that the differences in the R&D intensity of the service sectors in various industrialized countries are largely due to statistical peculiarities. But even if it were possible to get a better grip on the peculiarities of R&D statistics in the service sector, it cannot be expected that this would affect R&D's macroeconomic dimension.

In the wake of this situation, the political sector is assigned the task of formulating – in conjunction with suppliers and users of technologies, the science community and the economic sector, services and industry – guidelines and **objectives** for developing solutions that are marketable on global markets, albeit without prescribing the particular technological means to be used. Attention must be paid to fashioning the regulatory framework so that it is open to innovation and even fosters it. This applies above all to those areas which have been heavily regulated to date or which were originally the responsibility of the state. This would open up clear prospects for corporate innovation and research in these areas. It applies, for example, to the areas energy and the environment, education, health care and aging, leisure time and information, transport and mobility, nutrition, as well as to efforts to streamline government. To achieve this, the political sector would have to credibly prove its commitment to new technological trends, their problem-solving ability and the financial opportunities they entail – without veiling potential inherent risks. In this sense, "innovation policy" is both a cross-sectoral and a management task which plays the role of advocate for innovative solutions. Reviewing and pruning government regulations could help break through the stagnation in Germany's supply of services[35] and increase the intensity of innovation in the service sector.

---

[35]   Cf. also H.-H. Härtel and R. Jungnickel (1998), Strukturprobleme einer reifen Volkswirtschaft. Analyse des sektoralen Strukturwandels in Deutschland.

4 Indicators for Germany's Technological Performance

Another reason why increasing attention is being focused on the service sector's importance for innovation activity (Section 3) and, in the process, on its importance for an economy's technological performance is the fact that structural change throughout the world is presently geared to "tertiarization." Further, another global megatrend exists alongside the trend toward tertiarization – namely, the trend toward increasing know-how or "knowledge intensification." This trend waned somewhat in Germany – particularly in industry – during the first half of the 1990s. However, since coming out of its recession, Germany has fallen in line with it once again.[36]

The indicators on Germany's technological performance also center on the two cornerstones of knowledge intensification and tertiarization. Each of the subsequent sections examines these indicators using different time frames:

*   The ability to translate existing know-how into innovation and success in the marketplace is examined from a short-term perspective in Section 4.1.

*   The medium-term perspective applied in Section 4.2 is concerned primarily with industry's investment in new know-how, research and development and in real capital – in other words, with structural change that is foreseeable in the near future.

*   The long-term perspective taken in Section 4.3 focuses on long-term provisions for the future, and on needs and efforts in education and science.

4.1 The short-term perspective – industrial innovation and international technology markets in the midst of cyclical expansion

The German economy is currently on a growth track, albeit a flat one. The share of goods with high know-how content found in international visible trade is growing with each cyclical upswing. What is the German economy contributing in the present phase? Has technological know-how been translated to a sufficient degree and in sufficient time into inventions and innovation? Has Germany been able to maintain its position in international technology markets? What

---

[36] B. Gehrke and H. Grupp et al. (1995), Wissensintensive Wirtschaft und ressourcenschonende Technik.

consequences are emerging for industrial expansion and employment? To what extent does this trend spill over to the service sector?

Chart 4.1:   Triad patent of major industrialized countries

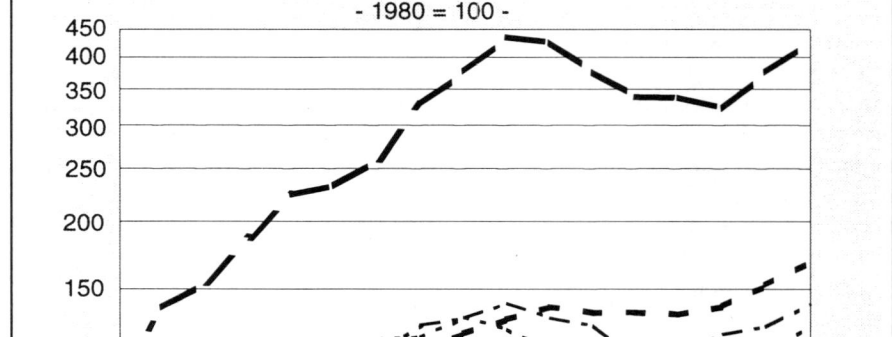

*   Former West Germany prior to 1990.
** Estimates for 1995 and 1996.
Sources: EPAT, INPADOC; FhG-ISI calculations.

4.1.1   Inventions

Inventions are the product of research and development. They represent the first stage in making (technological) use of know-how which comes primarily from industry. Patents provide a relatively accurate reflection of economically relevant invention activity. It must be remembered however that the importance of other forms of protection other than **patents** – such as copyrights – grows in tandem with the importance that services have for an economy's innovation activity.

Following a lull in the early 1990s, the number of "**triad patents**"[37] for products with global market potential began growing worldwide again in 1994. This trend accelerated further in 1996 (Chart 4.1). Without a doubt, momentum has returned to invention activity. However, it must also be assumed that part of this current surge in patent applications is due to growing international competition and the pressure it exerts to patent developments. Technological development is being driven by the search for technical solutions which can be utilized **quickly**. R&D projects are being examined more closely for their short-term exploitation potential. In this respect, it is not surprising that patent growth has developed considerably better in the short term than innovation and R&D efforts would otherwise seem to indicate.

Measured in terms of labor force and total population, Germany led the world's major economies in producing the most patents within the triad in the 1980s. Japan picked up considerable momentum during the second half of the 1980s and has been moving more or less in sync with Germany since 1990. Viewed in relative terms, Germany and Japan took top rankings with their number of triad patents in 1996 – ahead of the USA. However, the USA's patent standing looks different and more positive when gauged in terms of foreign trade orientation (export trade, for example). This is because patent protection is often sought specifically to increase export chances and to protect the import substitution sector against imitations.

Table 4.1:   Triad patents of selected countries 1996*

	Patent volume	Population in millions	Per million workers	Per PPP$ 1 bil. GDP	Per US$ 1 trillion exports	Per US$ 1 trillion exports + imports
Germany	8,499	104	216	4.9	16.2	8.6
France	2,652	45	104	2.2	9.2	4.7
Great Britain	2,943	51	103	2.7	11.2	5.4
USA	21,714	82	161	2.9	34.7	15.0
Japan	13,794	110	206	4.7	33.6	18.1

* Patent figures 1996 are projected.

Sources: EPAT, INPADOC; FhG-ISI calculations; Federal Statistical Office: Statistisches Jahrbuch für das Ausland; NIW calculations.

Germany is Europe's undisputed technological leader. Its triad patent intensity is double that reported for France and Great Britain. However, while the USA has been able to report an ongoing improvement in its position, the second half of the

---

[37]   This term refers to patents that are pending not only in the country of origin but also in at least **two foreign markets** in different regions of the USA-Europe-Japan triad. Inventions that are protected by "patents for products with global market potential" are typically of higher quality. Technology is not the only factor behind invention volume. Rather, the respective company's business policy – in other words, the degree and direction of its international focus – plays a role because, ultimately, foreign patents are acquired in order to obtain proprietary rights on the respective market.

1990s has seen Germany and Japan return to the position they held in the late 1980s. Great Britain and France also saw their patent volumes rise markedly in the last year under review.

Table 4.2:   Patent intensity (number of applications submitted to the EPO per million gainfully employed persons) for all technologies

Country	1989	1990	1991	1992	1993	1994	1995	1996
USA	135	140	138	138	139	143	159	175
Japan	209	205	180	163	162	155	181	206
Germany	417	386	293	299	303	322	342	410
Great Britain	144	131	128	128	129	133	139	153
France	213	204	205	193	197	204	212	233
Switzerland	543	522	445	483	466	483	492	550
Canada	48	43	44	47	49	51	59	68
Sweden	216	218	214	250	270	325	344	417
Italy	97	94	96	90	99	103	110	129
Netherlands	259	233	217	216	221	222	249	297

*   Figures for 1995 and 1996 are projected.

** Data prior to 1991 applies to former West Germany only, thereafter to Germany as a whole.

Sources: EPAT, PCTPAT; FhG-ISI calculations.

Germany's patent volume within the triad was quite stable throughout the 1980s but then slipped substantially during the early 1990s. German industry's greater "domestic orientation" was a contributing factor in this: The triad temporarily lost some of its importance as a sales market in the wake of German unification. The rather rapid increase observed in recent years was due not only to industry's increased international orientation but also to the fact that east German inventors have become more active (see also Section 5.2) and their focus on international markets has intensified.

Statistics on patent applications to the European Patent Office (EPO) reveal rather high levels for several smaller countries such as Switzerland, Sweden and the Netherlands. Apart from Germany, strong growth is reported by smaller countries in particular (such as Sweden and Canada). Even though they have little impact on the global economic trend, the gains these countries have made mean increased productivity (Table 4.2).

4.1.2   Innovation activity

The tendency to translate industrial know-how into marketable inventions more speedily can also be discerned in industry's innovation activities[38] which have picked up pace – albeit with a slight time-lag – alongside economic activity. Companies increased the amounts they budgeted for innovation activity in 1996/1997 (by 9% over 1995) – the first time since the 1992 (the last cyclical peak, for the time being). With the investment slump continuing through 1997, spending on innovation did not reach pre-recession levels in all areas (Chart 4.2). Although innovation-related investment in 1996/1997 rose by three to four percent, investment efforts still appear feeble in comparison to both 1992 levels and the increase that would normally be expected during a cyclical upturn. Only for 1998 are businesses planning to expand their innovation budgets substantially for the first time again, particularly in the area of investment.[39]

The number of firms that generate and develop innovation has risen markedly once again as well (Chart 4.3). The percentage of companies that spawn product or process innovations exceeded the 1992 level for the first time in 1997. Nearly 55 percent of Germany's industrial enterprises launched new products between 1995 and 1997, compared to 45 percent between 1991 and 1993 and 48 percent during the period 1993 to 1995. It is to be assumed that innovation activity and expenditure increased sharply in those areas in particular which typically invest very large amounts of effort and resources in R&D ("cutting-edge technologies"). This upward trend also continued through 1997. Industry is presently at efforts to win back the ground it lost during the first half of the 1990s. However, the recessionary trough that marked the first half of the 1990s provided companies the opportunity to substantially revise their innovation behavior.

---

[38]   Cf. regarding subsequent remarks and the definition of innovation and innovation expenditure, ZEW (1998), Innovationsverhalten im Verarbeitenden Gewerbe. Ergebnisse der Erhebung 1997. Innovation activities include all scientific, technical, commercial and financial steps taken to develop new or improved products, services or processes and launch them on the market. This enumeration also sheds light on the difference between innovation activities and research and development (R&D). In the international technology contest, R&D (see Section 4.2.1) is necessary on a long-term basis but is by no means enough in and of itself to improve international competitive strength. Other factors (such as the adaptability of new technologies, the creation of new products through design, and near-market innovation expenditure) have become more important for innovation capability and the utilization of growth potential. The term "innovation activities" is broader than "R&D."

[39]   Total expenditure – approximately a third of which is investment spending – is expected to increase by an estimated 10%. However, budgets for innovation activity in the industrial service sector (without wholesale and retail trade) were increased even more between 1996 and 1998: By 5% from 1996 to 1997 and by approximately 13% in 1998 (anticipations data).

Chart 4.2:   Industry expenditure on innovation

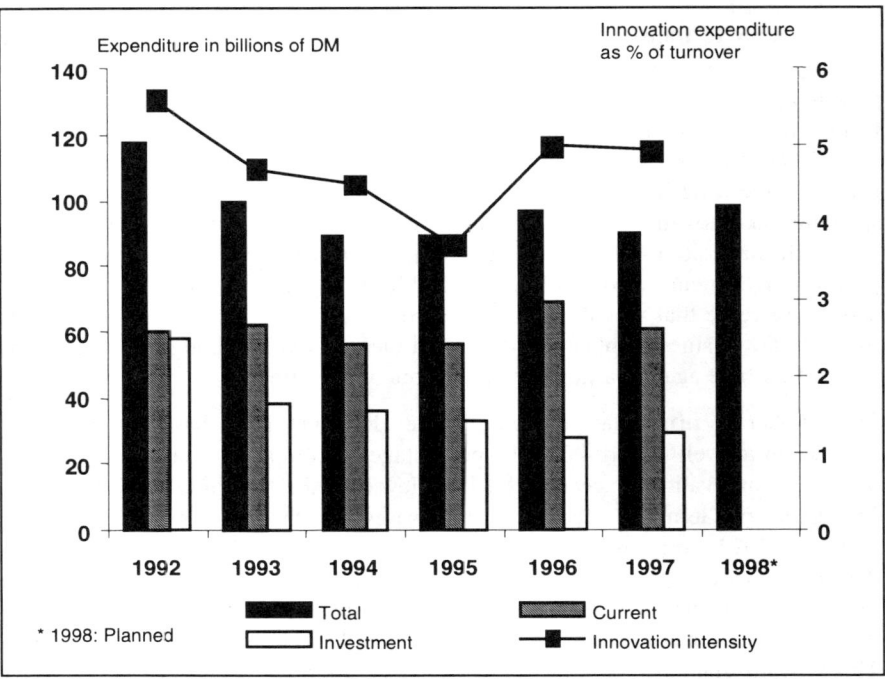

Source: ZEW Mannheimer Innovationspanel

- Sustained structural shifts are evident in innovation budgets. Increasingly smaller amounts are being allocated for investment expenditure while more and more is being spent on current operations: During the recession, **investment** spending was cut back faster than **current** expenditure on the short-term application of know-how – which was the first item to be raised with the coming of the upswing. By contrast, expenditure earmarked for **R&D personnel** and **R&D facilities** – which tends to be long-term in nature – has picked up pace only very slowly. In fact, it has increased at a slower rate than overall investment in fixed assets.

- The long-term trend reveals that R&D budgets have been cut (see Section 4.2.1) in favor of implementing know-how, conducting innovation activity and shortening innovation cycles. However, the trend toward conducting **R&D on a more continual basis** once again – albeit still along a low baseline – has manifested itself recently, particularly among small and medium-sized enterprises. Companies' R&D behavior and innovation behavior are usually very closely linked on a medium-term basis. In particular, a company frequently uses external know-how to complement its own R&D efforts. Collaborative projects are becoming increasingly important. The **ability** of small and medium-sized enterprises to work with research institutions and industrial companies on R&D **on a collaborative basis** increases to the same degree that their involvement in R&D continues at a stable level.

Chart 4.3:   Innovators in the industrial sector

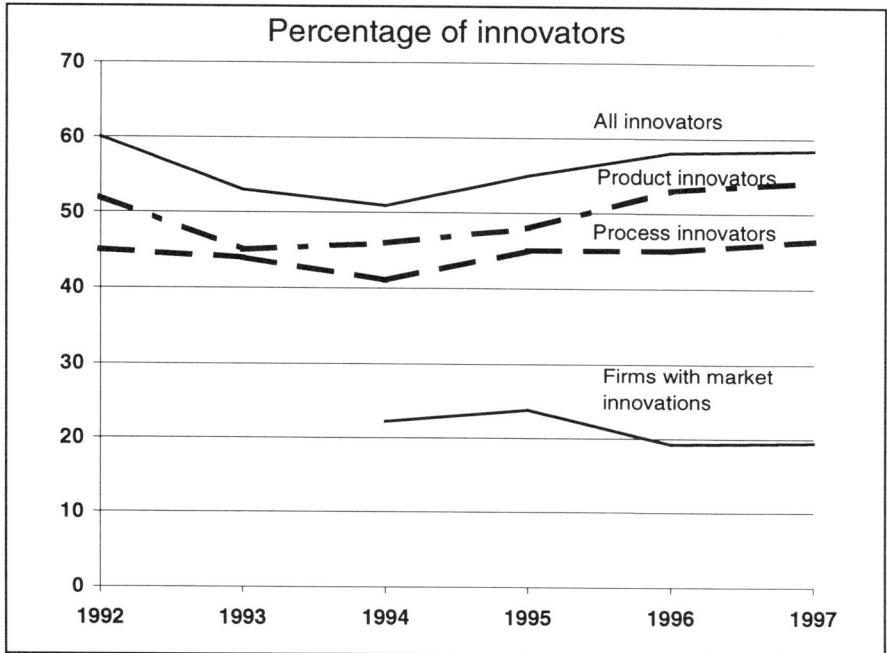

Source: ZEW Mannheimer Innovationspanel

* From an economic point of view, the speed at which inventions can be translated into marketable products and processes is just as important as an increase in R&D and a more rapid succession of inventions. Having "weeded out" their production programs during the recession, companies with **product innovations** to offer saw their share of turnover expand appreciably during the upswing. Today, innovation increasingly targets new, improved products (Chart 4.3) and modern **production programs**. The fact that new products account for a growing share of turnover[40] indicates however that product life cycles are becoming ever shorter and that competition over product innovation has heated up.

* In addition, manufacturing companies that produce innovation are also employing fewer people than in years past.[41] The percentage of companies that

---

40   A change in the wording of the question makes it impossible to directly compare figures for 1996 and later years with figures from previous years.

41   Since the impact that innovation has on employment is largely indirect, it is not however possible to draw firm conclusions about the development of total activity

produce unadulterated **process innovations** has however decreased further. Straightforward rationalization measures without product innovations are seldom successful any more.

- The increased market-orientedness of innovation activities and the continuation of R&D efforts at stable levels have had at least a short-term effect. The percentage of turnover generated by companies with product or market innovations has risen once again. However, a drop was observed in the percentage of companies that were able to produce **market innovations** (Chart 4.3) – in other words, products or services that are innovations not only for the respective company but at macroeconomic level as well, making them an important indicator for an economy's structural change and innovation capability. Many **product innovations** tend to be either gradual refinements and developments of an existing product, product differentiations or imitations. The large share they represent indicates however that technical know-how is being diffused at an accelerated pace.

This examination focuses on **industry**'s innovation behavior against the backdrop of cyclical developments. However, the observation of innovation activity in the **service sector** should be intensified, particularly in regard to longer-term aspects. Innovation activity in the service sector is to be assessed more positively than innovation in the industrial sector, which underscores the fact that the potential that service innovation offers is increasingly important for the economy (for further details, see Section 3).

4.1.3   Export activity in R&D-intensive industries

Industry's R&D-intensive sector is subject to tough international competition.[42] Companies must hold their own vis-à-vis foreign providers on the domestic market, even through their research findings and the patents they lead to do not necessarily target export production. Since the German market offers only limited opportunities for expansion anymore, an ever-larger portion of innovation activity is being directed to developing new, growing markets outside of Germany and – increasingly – outside of Europe and in Central and Eastern Europe's countries in transition. Until recently, export trade was the sole driving force behind innovation and growth in the industrial sector.

German foreign trade in R&D-intensive goods exhibited markedly greater momentum than total trade in manufactured goods. Germany exported some DM

---

from this. In other words, the impact that innovation has on employment is incurred among "users" of innovation, particularly in the expanding service sector.

[42]   Export earnings account for an average of nearly 50% of turnover in industry's R&D-intensive sector, compared to a good 20% in the non-R&D-intensive field.

370 billion in R&D-intensive goods in 1996.[43] This figure represents half the country's total manufactured goods exports for the year. At DM 230 billion, imports of R&D-intensive goods accounted for nearly 40 percent of Germany's total imported goods.

Chart 4.4:   World trade shares held by Germany, the USA and Japan in R&D-intensive goods 1989 – 1997/1998*

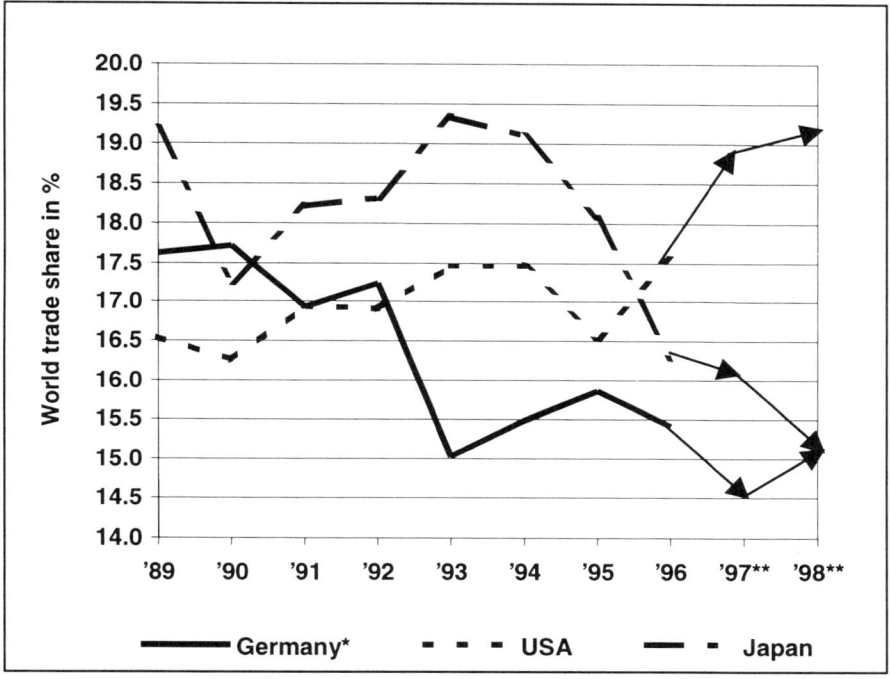

1. Data for the years 1989 through 1994 was calculated on the basis of the OECD-29.

*   Figures are for Germany as a whole starting 1991 and are therefore only partially comparable to previous years.

** Rough estimate.

Sources: OECD: Foreign Trade By Commodities; unpublished data for 1989 - 1995, 1995 CD-ROM; Federal Statistical Office; NIW calculations and estimates.

Germany ranks third among western industrialized countries as an exporter of R&D-intensive goods with a **world trade share** of nearly 15.5 percent, following the USA (nearly 18%) and Japan (16%) and considerably ahead of Great Britain (not quite 8%), France (7%), Italy (5%), Canada and Belgium (some 4% each) and

---

43   The most recent nuanced import and export data available are for the year 1996. Corresponding data for 1994 place exports at DM 320 billion and imports at DM 200 billion.

South Korea and the Netherlands (approximately 3.5% each). Nevertheless, Germany's world trade share in 1996 was smaller than in 1995 (Chart 4.4), primarily as a result of the weaker Deutschmark. However, even when data has been adjusted for fluctuations in the exchange rate and is viewed on a longer-term basis, Germany has yet to regain the ground that it lost on the global market during the early 1990s. Germany's world trade share might have improved slightly **in real terms** in 1997 because, measured in terms of volume, the country's export trade in advanced manufactured goods was heading for another record.[44]

Demand from **abroad** accounted for 90 percent of the **revenue growth** reported by Germany's R&D-intensive industries between 1995 and 1997, and has been the decisive driving force behind the upswing since late 1993. It continues to fulfil this function, having picked up further speed in 1997 (Table A.4) and then spread to non-R&D-intensive areas in the course of 1998. This momentum varies considerably: R&D-intensive industries which usually spend a very large share of sales on R&D (**cutting-edge technologies**) reported an average annual growth rate of 17.5 percent. This figure was more than 20 percent for 1996/1997 alone. Industries where R&D intensity is above average but not as high as in cutting-edge technologies (**advanced technologies**) increased their foreign sales by 9.5 percent a year.

4.1.4   Production and employment in the R&D-intensive sector

Production

Other countries expect predominantly high-quality, R&D-intensive goods from Germany. Driven by demand from abroad, R&D-intensive industries have once again been setting the pace for industrial production since the mid-1990s in a revival of a similar trend that marked the 1980s when they shaped industrial growth. Although German industry's R&D-intensive sector suffered above-average inroads into its growth rates during the previous recession, it was able to start picking up speed sooner and is likely to have exceeded its pre-recession production level for the first time again in 1998. However, those sectors in which R&D is not as highly ranked have not yet gotten back on track. The current economic expansion will probably continue through 1999 at a somewhat dampened pace – with the technology-oriented sector reporting further share gains.

---

[44]   However, the Deutschmark's markedly lower exchange rate vis-à-vis the US dollar must be taken into account. Calculated using the respective exchange rate, Germany's **nominal** world trade share probably shrank.

Chart 4.5:   Net output of R&D-intensive industrial sectors in Germany 1991 – 1998

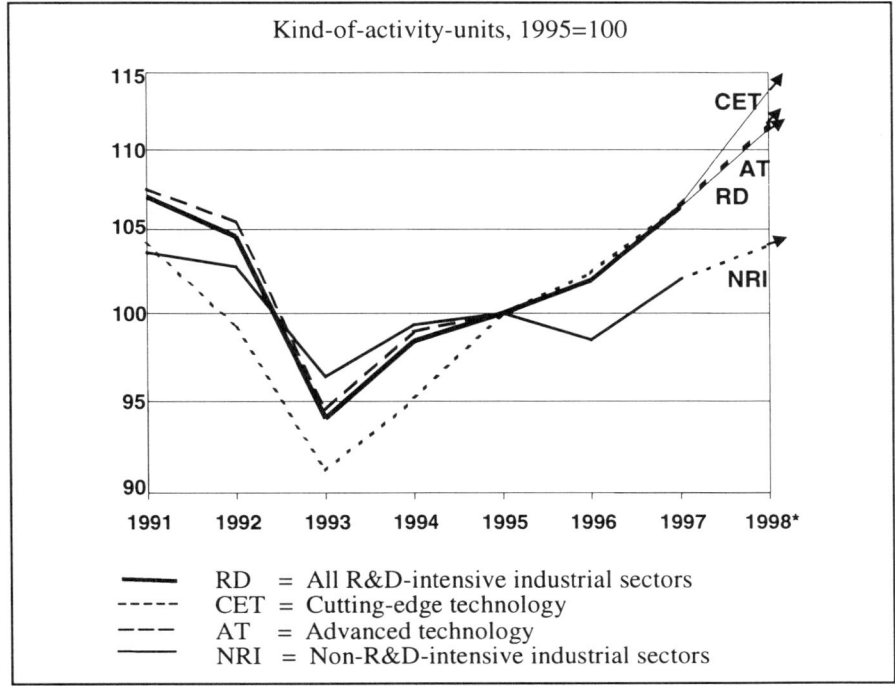

Kind-of-activity-units, 1995=100

RD = All R&D-intensive industrial sectors
CET = Cutting-edge technology
AT = Advanced technology
NRI = Non-R&D-intensive industrial sectors

* Rough estimates.

Sources: Federal Statistical Office: statistics on the manufacturing industry; NIW estimates
and calculations.

The R&D-intensive sector is undergoing structural change in the course of this growth process: Cutting-edge technologies have grown rapidly (with production increasing by 4% on annual average) and now top the growth list with approximately 7.5 percent of industrial value added (Chart 4.5). New, basic technologies that generate growth opportunities are being developed in these areas. Other industries, including the advanced technology field (primarily capital goods and intermediate producer goods), are taking up these opportunities and are characteristically starting to pick up speed only in the later phases of the economic upswing (3% a year[45]). WestLB expects production in R&D-intensive industries to expand by eight percent in 1998, compared to not quite four percent in all other industries.

---

[45]   During comparable upward phases in the 1980s, cutting-edge technologies in **West Germany** reported 8% production growth on annual average; this figure was nearly 3.5% for advanced technologies.

In **cutting-edge technologies**, the telecommunications field in particular has proven to be the source of "robust" growth, improving its position even during the recession.[46] The – albeit rather narrow-based – basic pharmaceutical products branch was also able to report constant growth. On the other hand, other areas of cutting-edge technology which currently buoy growth (electronic components, data processing and pesticides) have certainly not been immune to recessionary slumps. Otherwise, the economic upswing of the 1990s has clearly diverged from – and in some cases even inverted – the growth hierarchies established in the 1980s. The weapons/ammunition and measuring and control engineering fields have exhibited weak growth, and signs of structural problems are becoming apparent in the pharmaceuticals field and, even more so, in the aircraft industry, both of which have reported declining production for a long time, even during phases of economic expansion.

Not all industries in the **advanced technology** field will be able to resume the roles they played or the growth rates they achieved during the 1980s: This is due on the one hand to a shift in demand preferences and in the application of cutting-edge technologies: Namely, away from industrial enterprises and toward advanced services. On the other hand, price and cost play a larger role in defining competition between companies on advanced technology markets than on markets for cutting-edge technologies. International competition is toughest over these two factors. The advanced technology sector's overall production levels have only just reached the level reported in 1991. This sector's expansion is due primarily to the stable growth reported by the car industry and its suppliers in the chemical sector (plastics, paints and intermediate products) and individual specialized branches of the mechanical engineering field, advanced ceramic products, electric motors and photographical and optical products (for details, seeTable A.5).

Employment

The R&D-intensive sector employed 2.7 million – approximately 45 percent – of the six million people working in Germany's manufacturing industry in 1997. The fact that this sector generates some 50 percent of value added indicates that labor productivity in R&D-intensive industries is higher than average. The number of persons employed in R&D-intensive industries has fallen from year to year, and the correlation between employment and production growth has continued to weaken. Although the R&D-intensive sector has experienced relatively strong expansion, its employment situation is no better than that of other industrial fields where production is less research-intensive because the R&D-intensive sector is

---

[46]   But after Deutsche Telekom wrapped up its investment in its digital network, this sector was hit by a slump which had a particularly strong impact in 1997. However, new providers who are still using Deutsche Telekom's network are expected to have to invest in building their own networks soon.

under particularly strong pressure from international competition.[47] Only a few areas (the automobile industry and associated fields of electrical engineering; medical engineering; combustion engines/turbines) were able to produce a positive employment balance during the upswing. In addition, only the pharmaceutical and weapons/ammunition industries (Table A.6) reported a rise in employment in 1997.

Generally speaking, a revaluation of the R&D-intensive sector in Germany is called for. Although economic growth and increased employment were still linked with one another and the R&D-intensive sector generated nearly all new industrial jobs during the economic expansion of the 1980s,[48] it can no longer be expected that the nation's employment problems can be **directly** solved by a rapid expansion of its R&D-intensive industries.[49] Therefore, this development can only be viewed within the overall context: Although (physical) industrial production in Germany is on the rise once again, it is the service sector that is generating increasing value added and new jobs. For this reason, the service sector is essential to mitigating the country's employment problems. However, the R&D-intensive industrial sector's real importance for economic growth and employment tends to be **indirect** because this sector is home to a substantial portion of a country's scientific-technological problem-solving expertise. Technology supplies solutions which the service sector puts to use with the ancillary effect of creating new jobs.[50]

4.1.5   Employment in know-how-intensive services

The portion of the work force employed in R&D-intensive industries has dwindled in recent years not only in Germany but in most other highly developed countries as well. This development is by no means just a cyclical reflex: It is related to sustained trends arising from structural change. The service sector's **know-how intensive fields**[51] are claiming growing shares of national production capacity – in

---

47   This is evidenced by the sector's import and export quotas which are quite high in comparison to those reported by the less R&D-intensive sector.

48   Cf. H. Grupp and H. Legler (1992), Innovationspotential und Hochtechnologie. Technologische Position Deutschlands im internationalen Wettbewerb.

49   A medium-term projection issued by Prognos AG (Deutschland Report No. 2, Basel 1998) also arrives at higher production growth rates for R&D-intensive industries (3% a year until the year 2005) than for non-R&D-intensive industries (not quite 2%). Jobs will however be shed at a faster rate in the R&D-intensive sector – by a total of 12% as opposed to 9% in non-R&D-intensive industries.

50   Cf. DIW, contributions to the 1995 report on Germany's Technological Performance.

51   Although the number of R&D employees is not as adequate an indicator for innovation capability in the service sector as it is for innovation capability in industry (see Section 3), the percentage of highly qualified employees does provide a relatively good approximation of the degree of a company's innovation capability,

Germany as well (Chart 4.6). This is particularly evident in employment trends. In Germany, know-how-intensive areas of the service sector employed some 8.6 million workers (who are liable to social security) in 1997. In other words, at least two-thirds of all jobs in Germany's service sector are to be found in know-how-intensive fields.[52] These fields account for 35 percent of all jobs in the manufacturing sector.

Although the trend toward tertiarization has been ongoing over the **long term**, it has not made much headway in recent years.[53] Further, the **current** growth reported by know-how-intensive service fields in Germany is not entirely in step with the long-term momentum. However, employment in these industries grew by 0.5 percent in the former West Germany in 1997.

Business services have been especially able to produce a positive employment balance in recent years by using and disseminating I&C technologies. Industrial demand in particular has grown markedly since the late 1970s. Viewed over the longer term, economic growth was strongest in the areas of legal counseling/management consultancy, commercial advertising and the private exhibition business. Additional jobs were also generated in, *inter alia*, the media field, the education sector, travel agencies, and the health care sector – in other words, often in government-regulated areas.

By contrast, employment is currently on the decline in the banking and insurance industries due to rationalization efforts which center on increasing the use of I&C technologies. This trend is likely to continue and will probably hit low-skill jobs the hardest. This also explains job cuts in other areas, such as typing services. The list of major losers is currently headed by the railway, postal services and – due to the current situation in the construction sector – architects' offices and engineering consultancies. Jobs have also been shed throughout the know-how-intensive areas of the transport and telecommunications sector. The supply and provision of services in Germany has been stagnating for some years.[54]

---

regardless of whether the company belongs to the industrial or the service sector. Cf. List 3 in the Appendix for a list of know-how-intensive sectors.

[52] This figure is low because it does not include self-employed persons, people in the professions or civil servants.

[53] This applies to both sectoral employment patterns and occupational trends among the work force. Cf. H.-H. Härtel and R. Jungnickel (1998), Strukturprobleme einer reifen Volkswirtschaft. Analyse des sektoralen Strukturwandels in Deutschland.

[54] The HWWA has examined this problem in greater detail (see H.-H. Härtel und R. Jungnickel, 1998, Strukturprobleme einer reifen Volkswirtschaft. Analyse des sektoralen Strukturwandels in Deutschland).

Chart 4.6:   Employment in know-how-intensive sectors in former West Germany
                  1980–1997

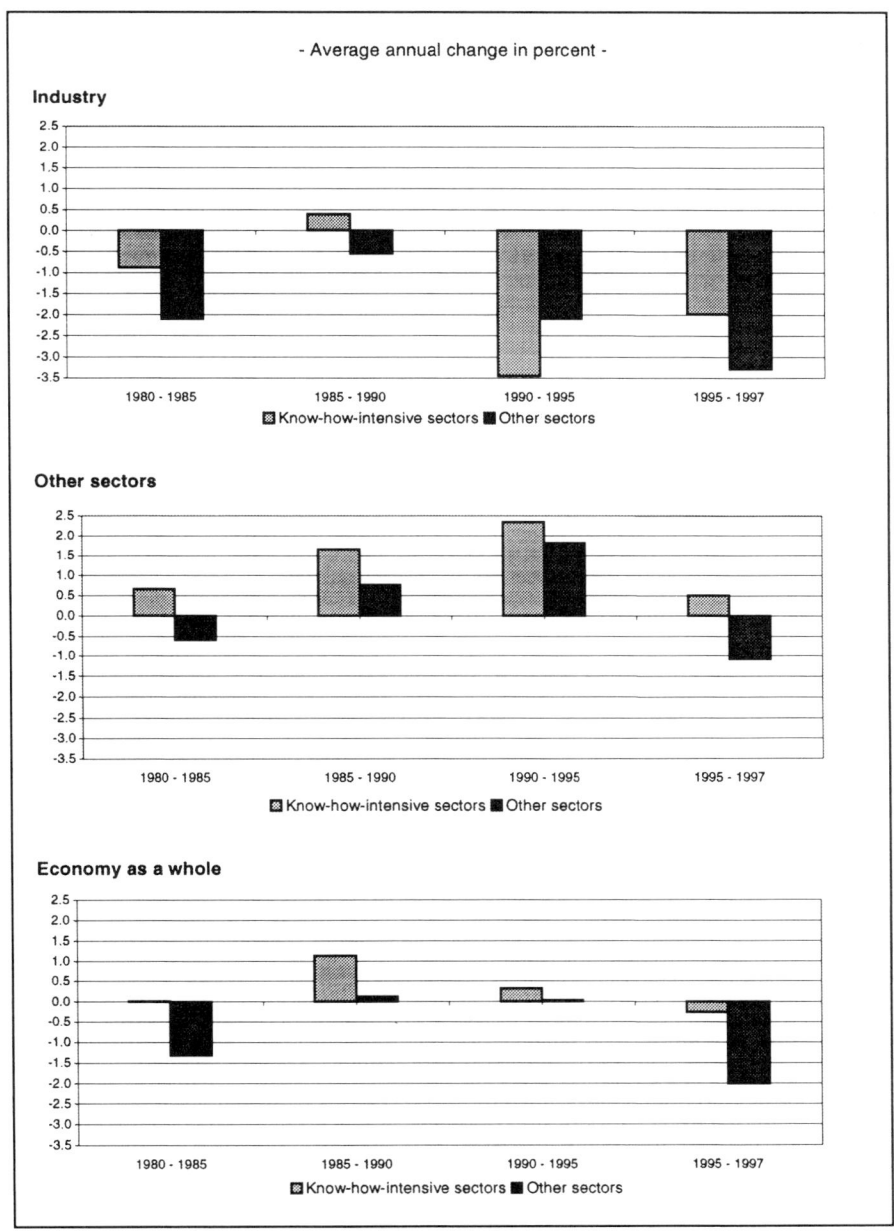

Sources: Federal Statistical Office: statistics on employees who are liable to social
         security; NIW calculations.

4.2   The medium-term perspective – investment in the near future

The question of how Germany's provisions for the medium-term future look arises particularly during an upswing. Are the opportunities being offered by the current export-driven surge in demand inducing new investment in research, development and innovation projects and for the development of new products and services that will stimulate the next wave of growth? Are the current upswing and existing innovation incentives strong enough to bring about a substantial expansion of productive capacity again? Are enough technology-based and know-how-oriented companies being started in the wake of structural change?

4.2.1   Research and development

Technical progress is based on research and development. Even in R&D-intensive industries, R&D reflects only a part – including however the "essential core" – of all innovation activity. Research and development is an investment in technological know-how which can be translated into products in subsequent years. This type of investment pays out much later than investments in fixed assets do. In this regard, R&D expenditure also reflects a company's assessment of its future prospects, and its willingness to undertake structural change. For this reason, R&D constitutes an important factor in determining technological performance. Particularly in the industrial sector, technological R&D is crucial for innovation activity. However, there has been a shift in weight in favor of services, both with regard to innovation activity and total economic activity (see Section 3.1).

4.2.1.1   *Research and development in international comparison*

Trends

Germany is one of the world's most research-intensive countries: OECD countries spent a total of US $460 billion on R&D in 1996. Some 42 percent of this amount was expended by the USA, 18 percent by Japan and 8.5 percent by Germany. Measured in percentage of the respective country's gross domestic product, Sweden tops the R&D-expenditure list, followed by Japan, South Korea[55] and Finland, Switzerland and the USA. At 2.4 percent (1997/1998), Germany is on approximately equal footing with France. Although Ger- many was to be found in

---

[55]   Although newly industrialized countries are making large-scale R&D efforts, this does not mean that these countries' technological know-how is presently anywhere near the level of know-how found in industrialized countries.

the front ranks in the early 1990s, it has since fallen back to take its place among the lower third of the current leaders, albeit still ahead of the Nether- lands and Denmark which have recently overtaken Great Britain (Chart 4.7).

Germany[56] was not the only country whose propensity to conduct R&D waned during the first half of the 1990s. R&D efforts in most major economies have remained stable or have been reduced in real terms. However, no other country scaled down its R&D as rapidly or as vigorously as Germany did. Although industry's propensity to conduct R&D[57] has declined in many major industrialized countries, private R&D efforts have in part been substantially increased in several smaller economies. This has brought a thorough reshuffling of the frontrunners among the world's most research-intensive economies. New entries in this list are Finland, Denmark and South Korea. Outside the top-ranking group, Canada and Ireland have been increasing their R&D efforts on an ongoing basis. Often the home base for multinational firms, these smaller economies focus mainly on selected fields of cutting-edge technology (telecommunications, I&C, pharmaceutical products/biotechnology) with correspondingly high R&D requirements as they work to catch up with the world's frontrunners.

The corporate sector in many larger countries has since sailed out of the phase of slack R&D activity that marked the first half of the 1990s. Investment in new know-how is on the rise once again. A turnabout in the downward trend has been particularly evident in the USA where R&D activity in the industrial sector has been growing by a substantial five percent a year since 1995 (Chart 4.8). Japan's industry has also stepped up its R&D efforts once again. By contrast, German companies have continued to practice R&D restraint. Only very recently has Germany also seen a change in direction toward more R&D. Consequently, Germany had been lagging behind for two to three years while international competition increased and other highly developed economies stepped up their R&D efforts.

The globalization of R&D

Even though the crisis in Asia will act to brake the globalization of R&D, efforts on the part of multinational companies to expand and restructure their global R&D

---

[56]   Germany's lower standing in R&D is due – in mathematical terms – only in small part to German reunification because the cutback in industrial R&D capacities initially progressed rapidly in western Germany as well. German unification probably had a decisive economic impact in the early 1990s because it generated strong demand in consumer fields. At the same time, technology-intensive capital goods industries were hit by the global recession which correspondingly affected investment propensity in Germany.

[57]   Most R&D is conducted by industry. Percentages range between 60% (in France) and not quite 75% (in the USA).

activities are continuing, particularly in R&D-intensive industries. The globalization of R&D activities has however centered on North America and Europe to date. Multinational companies are on the look-out first and foremost for markets for new, advanced products and services, and they look for them particularly in those areas which offer an assemblage of specific types of expertise and know-how which they can use to broaden their own R&D base.

Chart 4.7:   R&D intensity in selected OECD countries 1981–1997*

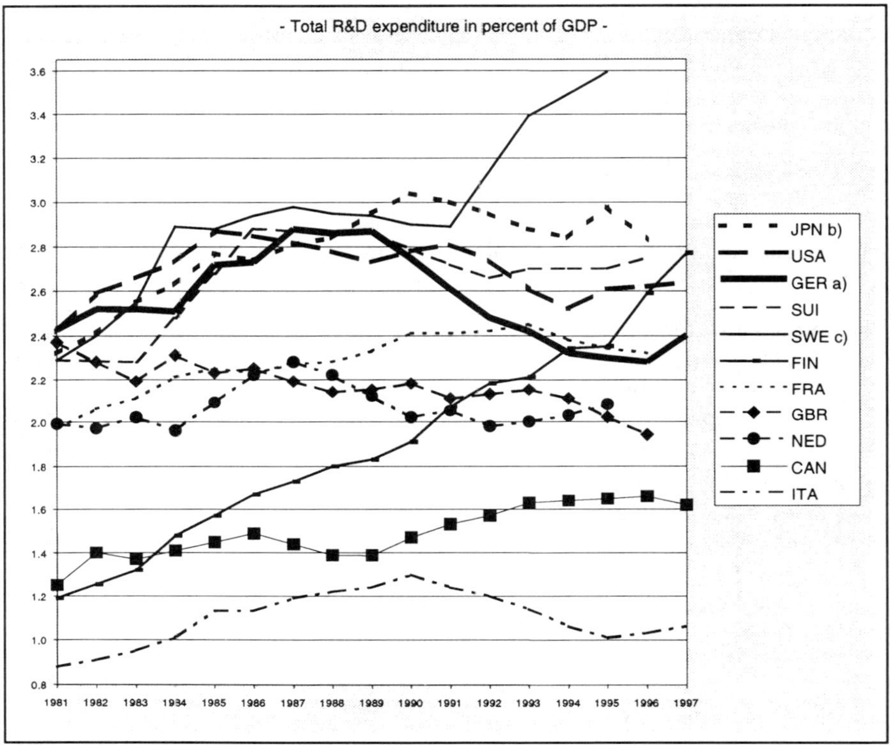

- Total R&D expenditure in percent of GDP -

(legend: JPN b), USA, GER a), SUI, SWE c), FIN, FRA, GBR, NED, CAN, ITA)

* Some of the figures are estimated.

a) Former West Germany prior to 1990.

b) R&D expenditure for Japan slightly overestimated up to 1995.

c) Break in sequence 1993/1995.

Sources: OECD: Main Science And Technology Indicators; NIW calculations and estimates; adjustments and recalculations for Germany by the BMBF.

A comparison of "host countries" places Germany in midfield. As a rule, foreign-owned companies and domestic companies exhibit similar levels of R&D intensity. Investment on the part of foreign-owned companies has generally not been seen to increase R&D activity. In the USA, the share of industry's R&D spending attributable to foreign companies has grown in recent years; many companies that conduct research have been sold to foreign owners. The growth in foreign R&D spending in the USA – which has slowed somewhat in the meantime – can be attributed to the country's attractiveness as a sales market and as a location for production and research. The large share of R&D capacity that foreign companies account for in Great Britain is due solely to the large share of foreign-controlled firms there. However, the decline in British R&D spending that lasted throughout 1996 as well casts a shadow on Great Britain's technological performance and shows that being attractive to foreign investors does not necessarily lead to increasing R&D expenditure.

Chart 4.8:   Industrial R&D intensity in selected OECD countries*

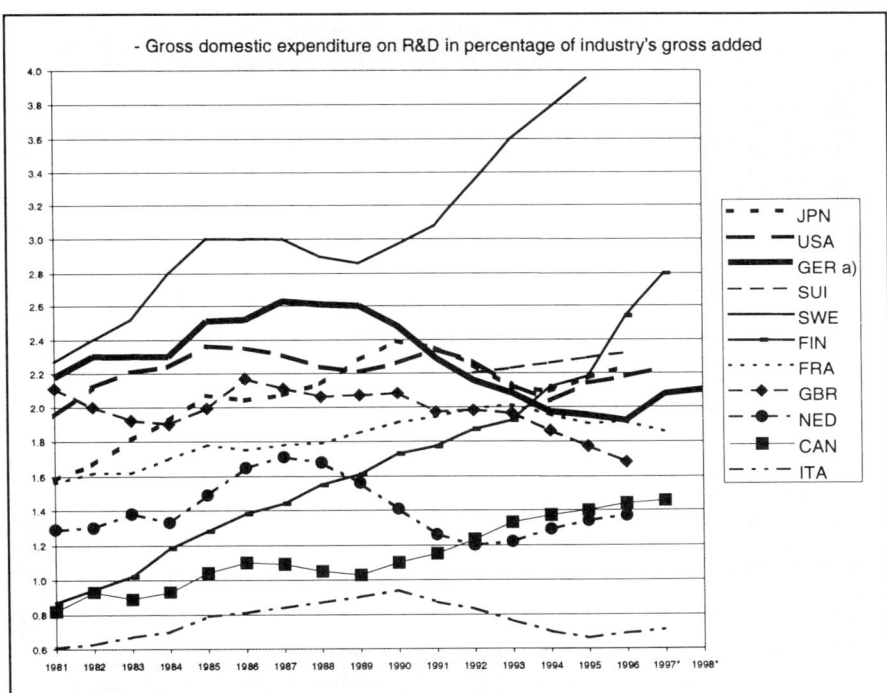

* Estimated.

a) Former West Germany prior to 1990.

Sources: OECD: Main Science And Technology Indicators; NIW calculations and estimates.

Government involvement

Viewed at international level, the downward adjustment of R&D activities in the industrial sector has probably bottomed out by now, whereas the **relative** decline in government involvement in R&D is likely to continue in most larger industrialized countries.

The type and degree of political intervention in the scope and thrust of technological developments varies from country to country. This is particularly apparent when Europe's three major economies Germany, Great Britain and France are compared with Japan and the USA (Table 4.3). Government involvement consistently accounts for nearly 37 percent of all R&D spending in Germany, twice the amount as in Japan. Despite the downward trend, more than 40 percent of all expenditure on R&D in France and nearly 33 percent in Great Britain is linked to government involvement. The USA in particular has seen a continual decline in the importance of government activities since the early 1990s – albeit starting from a lower baseline than in France; government involvement now accounts for less than one third of all R&D spending in the USA. Mixed trends can be observed in Japan where government involvement in R&D activities has been traditionally and comparatively restrained: This involvement was initially stepped up during the early 1990s and subsequently cut back during the second half of the decade.

Germany and Japan have certain structural features in common, albeit with considerable differences in level. In both countries, up to 90 percent or more of government-funded R&D is geared to civilian fields. By contrast, military objectives[58] account for a much greater portion of government-funded research in France (nearly 30%), and especially in Great Britain (almost 40%) and the USA (55%). Moreover, government-funded civilian research efforts in both Japan and Germany focus primarily on fostering basic research in universities (whereby Japan has most recently exhibited a pronounced trend toward cutbacks in this area) and on R&D programs aimed at structural improvement. In the USA, the government's research funding policy has assigned increasing priority to the health care and environmental fields for years now. The other reference countries – with the exception of Great Britain – compare badly with this. In addition, government efforts to promote technology in the USA and France have traditionally given high priority to space programs. This is a further characteristic of the American innovation system: By giving priority to the military field, health care and aerospace, it almost exclusively benefits very R&D-intensive industries.

---

[58]   The amount that the US government spends on R&D that is conducted for military purposes is some 50% greater than German industry's total domestic R&D expenditure.

4.2.1.2   R&D in German industry

Current developments

Although German industry has steadily shed R&D personnel for nearly a decade, cut-backs in R&D personnel have progressed at a somewhat slower pace than among other employees (Chart 4.9), particularly in research-intensive sectors.[59] For this reason, **R&D intensity** has once again increased slightly in recent years. However, only the automobile industry has actually increased its R&D capacities in Germany in **absolute terms**.

R&D expenditure was increased across a broad base in 1997, placing it ten percent above the level reported for 1995. As a result, R&D growth outpaced revenue growth for the first time in the 1990s. The R&D work force also grew for the first time since 1987, expanding by three percent over its 1995 level. While large-scale firms have buoyed this positive trend, an evaluation of small and medium-sized enterprises reveals them to be somewhat more subdued in their R&D spending behavior. However, it is already quite an achievement that small and medium-sized enterprises have abandoned the trend toward withdrawing from R&D in favor of conducting R&D activity on an ongoing basis. A glance at individual industries reveals the following picture: The automobile industry once again increased its R&D activities substantially, in excess of the average reported for the entire industrial sector. The chemical/pharmaceutical industry and the office machines/EDP, electrical engineering, and precision tools/optical goods/watches and clocks fields also reported sizable increases.

During the 1990s, industrial R&D was geared more closely to market anticipations and sales prospects than in previous years. Industry reacted very sensitively to **cyclical** influences and cut back its strategic research activities in particular.[60] Looking at benchmark data and estimates (the only data available at this time), it is not entirely clear whether the recent step-up in R&D efforts on the part of (major) firms has brought the cutbacks in strategic research to a standstill and, as a result, injected greater continuity into their R&D activities. Simply increasing industrial R&D capacity on a "cyclically neutral" basis is probably not enough to achieve this; the ongoing reduction in R&D levels since the late 1980s has been too sustained to allow this. In and of itself, an increasingly brighter economic outlook does not mean that there will be a trend toward a **substantial**

---

[59]   The number of job cuts was particularly large among R&D assistants and technicans, whereas the number of positions for scientists and engineers was reduced only slightly.

[60]   Cf. also the NIW's extensive report which was used for last year's report on Germany's Technological Performance. Licht, Schnell and Stahl (1996) and Licht and Stahl (1997).

improvement – particularly in light of the fact that companies in many other countries have markedly expanded their R&D efforts once again. German industry is straggling behind.

Table 4.3:   Government involvement in R&D* in G5 countries 1991–1998

	1991	1994	1996	1997	1998
	- Percentage of total / Percentage breakdown by individual				
GER					
Total	35.8	37.1	37.0		
Thereof: Civilian	89.0	91.4	90.2		
Thereof: Development	25.5	22.6	23.1		
Health/Environment	13.0	13.4	12.7		
Aeronautics	6.0	6.0	5.5		
Non-specific	17.0	15.8	16.5		
Basic research	37.3	41.4	41.3		
GBR					
Total	35.0	33.2	31.8		
Thereof: Civilian	56.1	61.1	62.4		
Thereof: Development	28.8	18.9	16.6		
Health/Environment	22.3	24.0	32.3		
Aeronautics	4.8	5.1	4.1		
Non-specific	9.1	19.3	18.6		
Basic research	33.7	32.0	27.8		
FRA					
Total	48.8	41.6	.	.	
Thereof: Civilian	63.9	66.9	70.3	72.3	
Thereof: Development	32.8	22.5	19.3	19.8	
Health/Environment	9.8	10.8	12.3	12.4	
Aeronautics	13.5	15.9	15.6	15.2	
Non-specific	23.9	26.6	26.6	26.5	
Basic research	19.4	21.4	22.8	22.8	
USA					
Total	38.7	37.1	33.6	31.6	
Thereof: Civilian	40.3	44.7	45.3	45.0	45.9
Thereof: Development	22.1	22.7	20.7	19.6	19.6
Health/Environment	43.5	44.2	45.1	46.7	46.7
Aeronautics	24.5	24.3	25.1	24.4	24.3
Non-specific	9.9	8.9	9.1	9.3	9.4
Basic research			n.a.		
JPN					
Total	18.2	21.5	18.7		
Thereof: Civilian	94.3	94.0	94.1	94.2	
Thereof: Development	33.5	31.4	34.4	34.8	
Health/Environment	5.7	6.2	6.9	7.3	
Aeronautics	7.2	8.0	7.0	6.7	
Non-specific	8.5	9.7	10.2	11.5	
Basic research	45.1	44.8	41.4	39.7	

GBAORD: Total government budget appropriations or outlays for R&D.

Sources: OECD: Main Science and Technology Indicators; BMBF data; NIW compilation.

Corporate plans however indicate that R&D expenditure is to be increased by some five percent in 1998. Predictions regarding the planning horizon in 1999 cannot be made with any certainty. Many more major companies than in the past plan to expand their R&D capacities; only a few firms plan to trim their R&D budgets. Ads for R&D personnel seem to confirm this. Which presages both the recruitment of young blood and the expansion of R&D staff sizes. The number of vacancies for natural scientists is growing, and has even doubled for data processing specialists. The "run" on vacancies is waning.

Impediments to innovation in the industrial sector

More innovation and R&D are urgently needed to buoy expansion and concomitantly help safeguard jobs. However, more than just technology is needed to generate innovation. Even with technical know-how expanding at a rapid pace, the speed of innovation can slow when conditions for putting know-how to practical use are unfavorable. For this reason, all conditions that are relevant to innovation must be examined closely to determine whether they stimulate or impede innovation.

A number of impediments reduce the anticipated return on the funds invested in an innovation project or directly increase the **market, cost and technical risks** it involves (Chart 4.10). **Financing** innovation projects also constitutes an obstacle, particularly for small and medium-sized enterprises. The negative effects of impediments increase in tandem with R&D intensity: As a rule, extremely innovative companies and industries with high R&D-intensity levels have more problems than non-R&D-intensive industries. Economic risk, lack of capital, and government regulation affect cutting-edge technology most.

Finding experienced, specially trained and highly-qualified **specialists** for an innovation project at short notice on the German labor market can be difficult. Because of such difficulties, projects often have to be extended. The automobile industry, for instance, reports a shortage of skilled labor.

Approval procedures and legal obstacles continue to be important impediments. They can lead to delays or even the discontinuation of an innovation project when they exceed the time frames the respective company has calculated for them. Regulations and lengthy administrative procedures impact the chemical industry to an above-average degree.

On the other hand, very few firms experience concrete problems with a lack of technical information or customer acceptance. Only companies in the plastic and rubber processing sector struggle to an above-average degree with a lack of **technical information**. This industry is an important customer of the research-intensive chemical industry which in turn laments **a lack of customer interest** more often than other industries do. Problems in conveying know-how apparently exist between the chemical industry and the plastic and rubber-processing industry

– in other words, between technology providers and technology buyers. It would appear that the chemical industry needs to improve the interface between its R&D departments and marketing departments.

Chart 4.9:   R&D expenditure as a percentage of gross valued added and R&D personnel as a percentage of the total work force in the manufacturing sector in former West Germany and in Germany as a whole

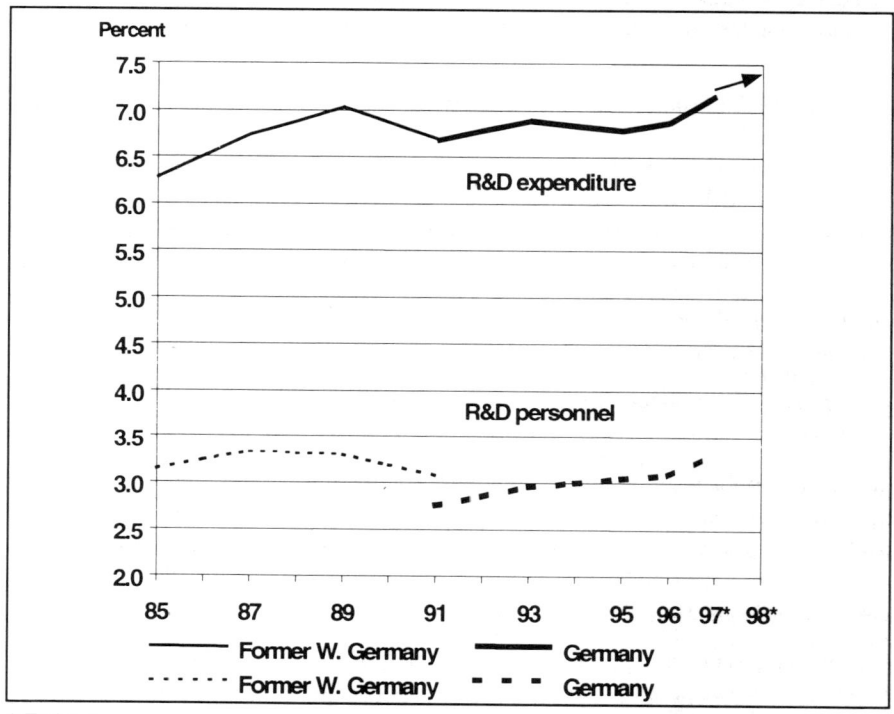

* Estimates.

Sources: SV-Wissenschaftsstatistik 1997; Federal Statistical Office: National Accounting System; OECD: Main Science and Technology Indicators; NIW estimates and calculations.

The internationalization of R&D

The research potential of German companies abroad grows in step with their investment, sales and production abroad. Prior to 1997, large German companies that had expanded their R&D capacity – when at all – did so primarily **abroad**. R&D spending by German companies abroad equaled some 17 percent of domestic R&D expenditure in 1997. This figure approaches 50 percent in the chemical industry. In fact, the chemical industry – followed by the electrical engineering and automobile industries – sets the pace for the internationalization of German industry's production and R&D activities. German car makers are in the process of rapidly catching up in the internationalization field, particularly vis-à-vis their American competition. Germany's leading automobile manufacturers are strengthening their research and production exposure on major markets via takeovers and the establishment of R&D facilities abroad. At the same time, they are the only industry that is steadily expanding its R&D capacity in Germany. The increasing globalization of this industry is apparently also strengthening its domestic R&D location.

Chart 4.10:   Companies directly affected by impediments 1994-1996

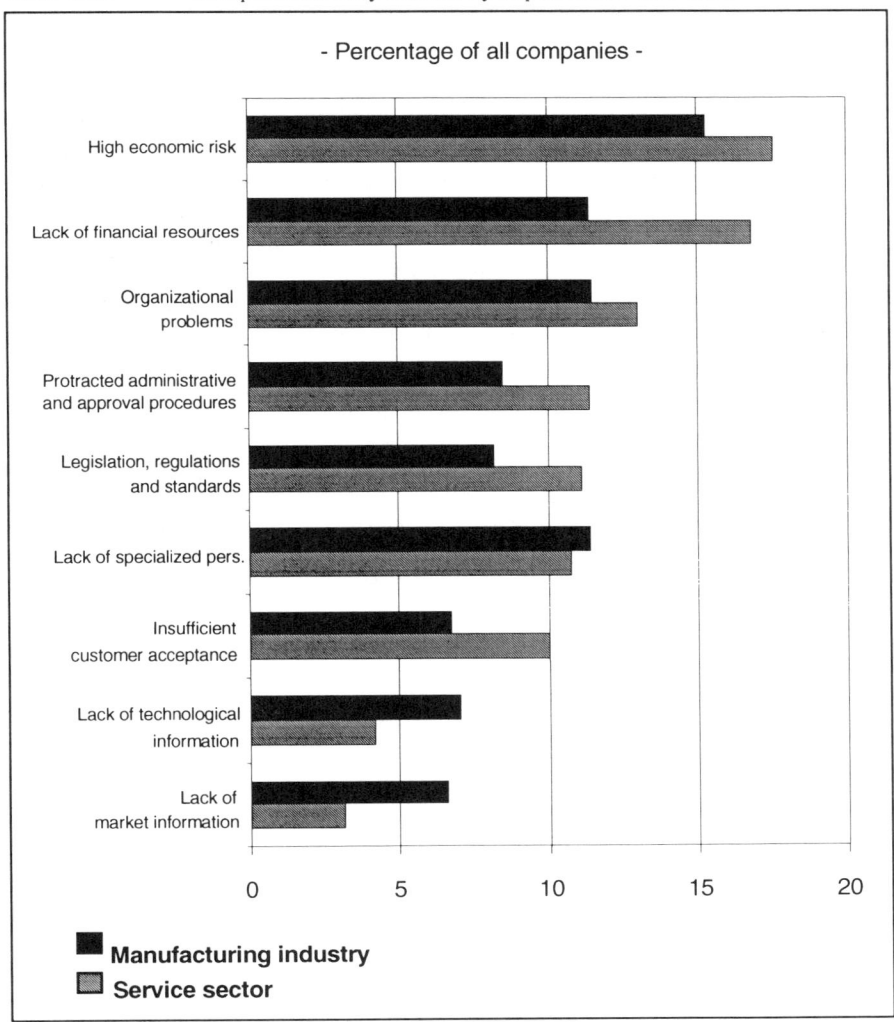

Source: ZEW/FhG-ISI: Mannheimer Innovationspanel.

The USA is the most important R&D location for German companies and accounts for more than half of the R&D funding German companies spend abroad. German firms spent approximately US$ 3 billion on R&D in the USA in 1996, compared to US$ 2.5 billion in 1994. As a result, they have the second largest foreign R&D capacity in the USA, following Swiss companies and ahead of British companies. The large increase in the amount that German companies invested in R&D activities in the USA during the 1990s can be attributed to major acquisitions and to the boom in corporate takeovers, particularly in the pharmaceutical industry.

In **Germany**, the internationalization of R&D is also defined by corporate takeovers on the part of foreign investors. This has however made little difference in the importance of foreign-controlled companies, which have fallen into line with the extended trend toward shedding industry jobs and reducing R&D capacity. At some US$ 3 billion, R&D expenditure on the part of US companies in 1996 remained unchanged over 1995. Accordingly, there are neither surpluses nor deficits in the "R&D balance of payments" between Germany and the USA. Although Germany's share of foreign research being conducted in the USA has diminished somewhat, it has topped the USA's list of preferred foreign research locations for quite some time, followed by Great Britain. Looking at the R&D intensity of US subsidiaries abroad, Germany ranked second behind Japan in 1996.

Assuming that foreign companies still tend to fall into line with the R&D behavior prevalent in the respective country, it can be expected that R&D expenditure by foreign companies in Germany – over and above the spectacular takeovers in the pharmaceutical industry – also began to rise again during the last two years.

4.2.2   Investment in capital goods in the R&D-intensive sector

Investment activity provides information about medium-term capacity planning, measures to modernize industry's production apparatus and, as a result, future technological performance.

The latest economic upswing has seen industry's R&D-intensive sector – driven by foreign demand – echo the role it played during the 1980s, namely, that of a driving force behind industrial growth: It has maintained the impetus it exhibited upon coming out of the last recession. The propensity toward investment among Germany's high-tech industries is (once again) growing (Chart 4.11).[61] Investment in structures and equipment equaled nearly 4.5 percent of turnover in 1996/1997, following four percent in both of the previous two years. Despite this, production capacity was expanded by only 1.5 percent a year following the recession, a rather

---

[61]   Investment activity in the former federal territory (i.e., West Germany) is shown in Chart 4.11.

meager rate.[62] Production capacity had yet to reach 1991 levels in 1997 (Chart 4.12). Substantial expansion is however to be expected according to the latest survey conducted by the ifo Institute on corporate planning for 1998. As a result, **nominal** investment expenditure will exceed 1989-1992 levels for the first time again in 1998. Looking at 1999, investment volume is not likely to grow as much as it did in 1998, and production growth will probably be only about half the rate reported for 1998.

Chart 4.11:   Gross capital investment in R&D-intensive industrial sectors 1989–1999

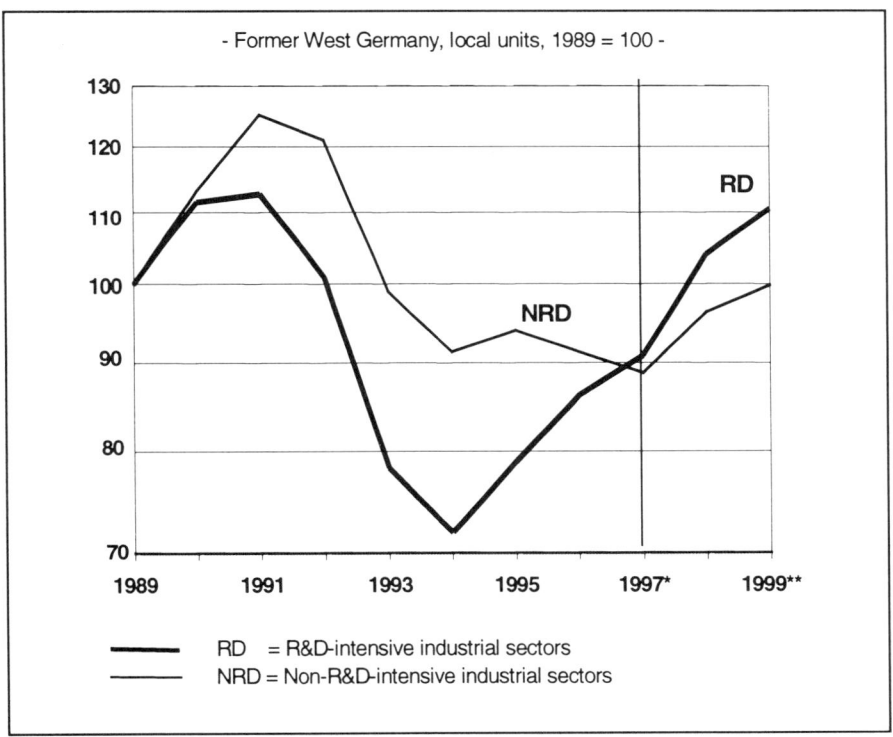

- Former West Germany, local units, 1989 = 100 -

```
RD  = R&D-intensive industrial sectors
NRD = Non-R&D-intensive industrial sectors
```

*   Estimates and

** Plans based on the Ifo Investment Test, autumn 1998.

Sources: Federal Statistical Office: statistics on the manufacturing industry; Ifo Investitionstest, autumn 1998; NIW calculations.

Viewed in structural terms, some 85 percent of the additional industrial capital expenditure that was spent between 1994 and 1999 (or is currently planned) has gone to the R&D-intensive sector. Industrial productive capacity is being geared at

---

62   Production capacity in industry's research-intensive sector grew nearly 3% a year during the period 1983 through 1990.

an increasingly rapid pace toward an intensification of research and know-how. The cutting-edge technology field and the automobile industry in particular are reporting large increases in investment once again. In some cases such as in the electronic component field however, these increases are due to investment in the construction of completely new production facilities in Germany's new *Länder* which has led to a sizeable expansion of this industry's production capacity. Other R&D-intensive industries have been able to get a firm foothold only in later phases (mechanical engineering, the aviation industry). Investment growth in the chemical industry is approximately on par with investment growth in industry as a whole, but is somewhat slower in the electrical engineering and precision instruments/optical products/watches and clocks fields. Most other – less R&D-intensive – branches of industry continue to stagnate.

As a reason for investment, **expansion of plant facilities** has only recently gained in importance, particularly in know-how-intensive industries. Rationalization/re-structuring and replacement were still equally ranked investment goals as recently as 1995. This develop-ment, together with the expansion of R&D capacity, would indicate that companies will be offering more technological innovation in their range of products in the future.

Chart 4.12:   Industrial production potential in former West Germany

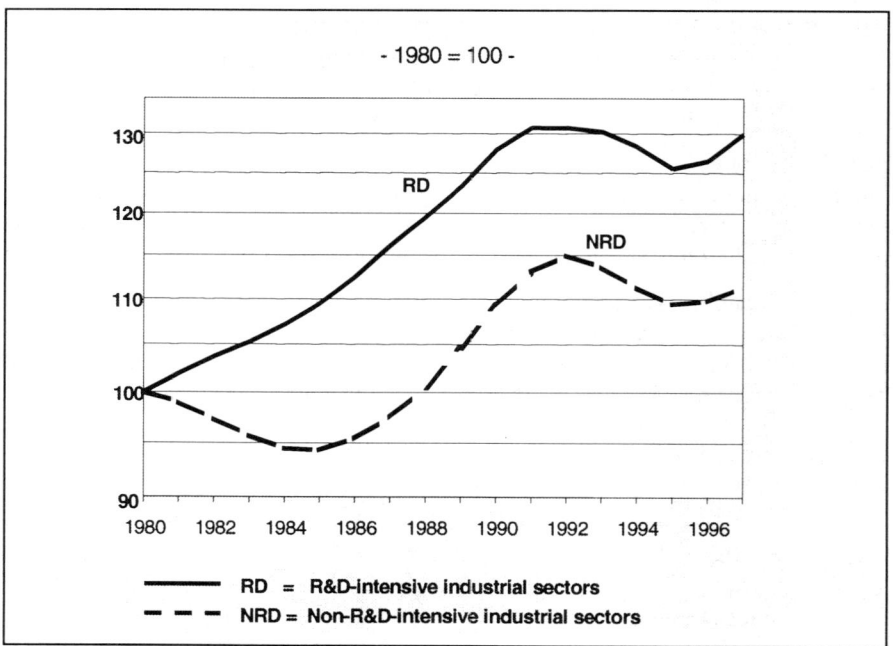

- 1980 = 100 -

RD = R&D-intensive industrial sectors
NRD = Non-R&D-intensive industrial sectors

Sources: DIW (Ed.): Produktion u. Faktoreinsatz nach Branchen des verarb. Gewerbes (October 1998); NIW compilations and calculations.

4.2.3   Company start-ups and closures

Germany's development into a knowledge-intensive society is proceeding not only from within established businesses but is also being pushed "from below" – in other words, by an ongoing "supply" of new companies which expand and modernize the range of products on offer with new business ideas, and thereby present a constant challenge to established firms. Young firms are an important prerequisite for keeping pace or catching up with international developments in "new" fields of technology and during the first stages of translating scientific findings into new products.

Even the large portion of new start-ups that vanish from the market after just a short time make an important contribution toward structural change. In such cases, either their business ideas or the potential innovations they "tested" did not pass the test of the market or they were taken over by an established firm or some other young company and presently hold their own in the marketplace in an improved form. They open up niches which established companies do not occupy or have not recognized.

It would however be unwise to hold unrealistic expectations because industries with high start-up rates also have high "mortality rates" among young firms. Nevertheless, when an industry's start-up and closure rates are about the same, young blood is rejuvenating the employers' ranks and, as a consequence, the respective industry is becoming increasingly competitive.

New businesses and self-employment

Looking at the entire economy, the number of start-ups in Germany grew during the 1990s, particularly in the know-how-oriented service sector (as was the case in the USA) where it does not take as long to become viable in the marketplace as it does for industry. While the total number of **new businesses** (*Neuerrichtungen*) increased once again in former West Germany in 1997, it continued to fall slightly in the new *Länder*. Parallel to this, the number of companies that closed in the 1990s rose sharply.

- The number of new businesses in **R&D-intensive industries** grew by 1.6 percent in 1997, markedly faster than the 0.3-percent growth in start-ups reported in non-R&D-intensive sub-sectors. Despite this, the share of start-ups in high-tech industries out of all start-ups continues to be rather small. Only recently has the pace begun to pick up substantially.

- Looking at the **service sector**, start-up activity has, as in past years, been progressing at an above-average rate in high-tech service industries (primarily in the software production field) and in other areas of business services, with the number of new businesses in these areas growing by approximately 6.5 percent in 1997.

A strong link exists between start-ups and their founders' educational training, with the likelihood that an individual will take up self-employment increasing in line with his level of education, particularly in the professions. Currently 19 percent of all university graduates and 13 percent of all technical college graduates are self-employed. The propensity to start up new businesses in (cutting-edge) technology fields is above average in the area along the "outskirts" of universities and non-university research centers. Further, when seeking a field of activity, university graduates look primarily to innovative areas of the service sector (where the share of self-employed persons grew from 16 to 23 percent during just the first half of the 1990s). Institutions of higher education could give further impetus to this trend by providing more business education.

Germany's **self-employment rate** has risen sharply in recent years due to two factors: Firstly, an accumulated need for services and, secondly, the declining number of persons in dependent employment and the resultant greater threat of unemployment.[63] The fact that the percentage of self-employed persons in the USA and Great Britain tends to be on the decline does not mean that the trend is negative in these countries. A look at the backdrop to this decline puts this picture into perspective: Both the number of self-employed persons and the number of persons in dependent employment rose – particularly in the USA – during the 1990s. On the other hand, the **number of self-employed persons** in Germany has increased only slightly while overall employment has **declined**. It must also be noted that this growth has been due to the increasing number of self-employed persons who have no employees (Table 4.4), whereas the number of self-employed persons who have employees has been on the decline since 1994. Business services and transport/storage/communication constitute an exception to this trend, with the number of self-employed persons with employees rising in both these areas in recent years.

All in all, self-employment has generated fewer and fewer jobs in recent years. Based on the number of persons who have just recently taken up self-employment, fewer jobs are being created today when an individual becomes self-employed than in the early 1990s.

Start-ups and closures

Even if all newly founded companies were to take up market positions already occupied by established companies, they would still have the effect of "rejuvenating" the sector. Given that a new company in the marketplace frequently generates additional innovation activity, this would have to be evaluated positively. However, the "crowding-out effect" is greater in those

---

[63]   So-called "tideover" allowances are granted to foster self-employment when unemployment would otherwise threaten.

industries where start-ups and closures offset each other because new start-ups generally generate fewer new jobs than are eliminated when a company closes.

- High-tech service sectors (Chart 4.13), cutting-edge sectors of technology, and business services report large "surpluses" of new start-ups over closures (albeit with a trend toward smaller surpluses). Such surpluses would indicate that these areas will continue to gain in importance. In this connection, competition over new ideas is particularly intensive in know-how-intensive branches of the service sector.

Table 4.4:   Self-employed persons: Total and self employed persons without employees

- Former West Germany -	Self-employed persons in thousands			Self-empl'd pers. w / o employees in percent		
Sector	1995	1996	1997	1995	1996	1997
Producing sector	608	632	644	32.9	38.0	37.7
Manufacturing	355	345	369	34.6	40.6	40.1
Construction	249	285	273	30.1	34.7	34.4
Trade, transport &	883	894	914	38.7	41.2	42.7
Trade, restaurants and	777	790	798	37.7	40.3	41.9
Transport, storage &	106	104	116	46.2	48.1	48.3
Total other services	1,012	1,086	1,155	51.9	55.6	57.4
Financial institutions &	87	87	91	62.1	65.5	65.9
Business services	400	428	479	51.8	54.4	57.0
Community, soc'l & pers.	525	571	584	50.3	54.8	56.5
Total industry	2,850	2,921	3,014	45.8	48.6	50.0

Sources: Federal Statistical Office: Fachserie 1, Reihe 4.1.1, various issues; ZEW calculations.

- On the other hand however, the advanced technology field reported a more or less equal number of start-ups and closures throughout the 1990s. The start-up surplus from the 1980s waned over time. Corporate growth in the advanced technology field and in other industrial sectors is restricted for the most part to company succession – in other words, the founding of new and closure of existing companies – and, consequently, to intrasectoral structural change and to an enhancement of competitive strength as a result of this "natural selection."

- The number of closures outnumbered the number of start-ups in all other industrial sectors during the 1990s. In other words, the number of companies declined over the decade. On the other hand, consumer services, wholesale and retail trade and the transport and telecommunications sectors exhibit a slight tendency toward expansion.

Start-up and closure rates are determined primarily by the market entry and exits costs specific to the particular industry and by the forces driving growth at a particular time. Medium-term market expectations are reflected in the rate of new start-ups, while short-term cyclical effects tend to affect the momentum behind market exits:

- The rise in the number of closures in the industrial sector should be interpreted as a reaction to the economic crisis of the 1990s.

- However, the sharp drop in the start-up rate in the cutting-edge technology field between 1989 and 1993 was due more to the generally unfavorable market anticipations prevailing at that time, and should be interpreted less as a structurally-related long-term decline.

4.3   The long-term perspective – education, science and research

In the long run, Germany's schools and universities will decide the question of whether German industry's international competitive strength will last – because education, science and research are the decisive factors in determining potential for growth, innovation and employment. Well-trained workers and scientists fuel greater output of know-how and contribute to the rapid dissemination of know-how. For this reason, qualified work and a high level of scientific research constitute **the** best cards that highly developed economies such as Germany have to play in the game of international competition over industrial locations. In the end, innovation is the fruit of investment in education and science. A lack of investment in education and science could eventually hobble innovation, growth and employment.

The current upswing offers an opportunity to expand the long-term foundation for German industry's technological performance – namely, education and science – more intensively than last observed. Rather than impairing technological performance immediately, adverse developments in education and research have an impact only after a time-lag of ten to 15 years. It is only with considerable effort and at great expense that such developments can be remedied. The level of investment in education and science provides the primary indication of whether enough is being done to secure future economic prosperity.

4.3.1   Trends in skill requirements

Any examination of the demands made of education and science also requires an examination of vocational requirements and long-term economic structure trends.

Chart 4.13:   Start-up and closure rates in selected economic fields in former West
Germany (including West Berlin)[64]

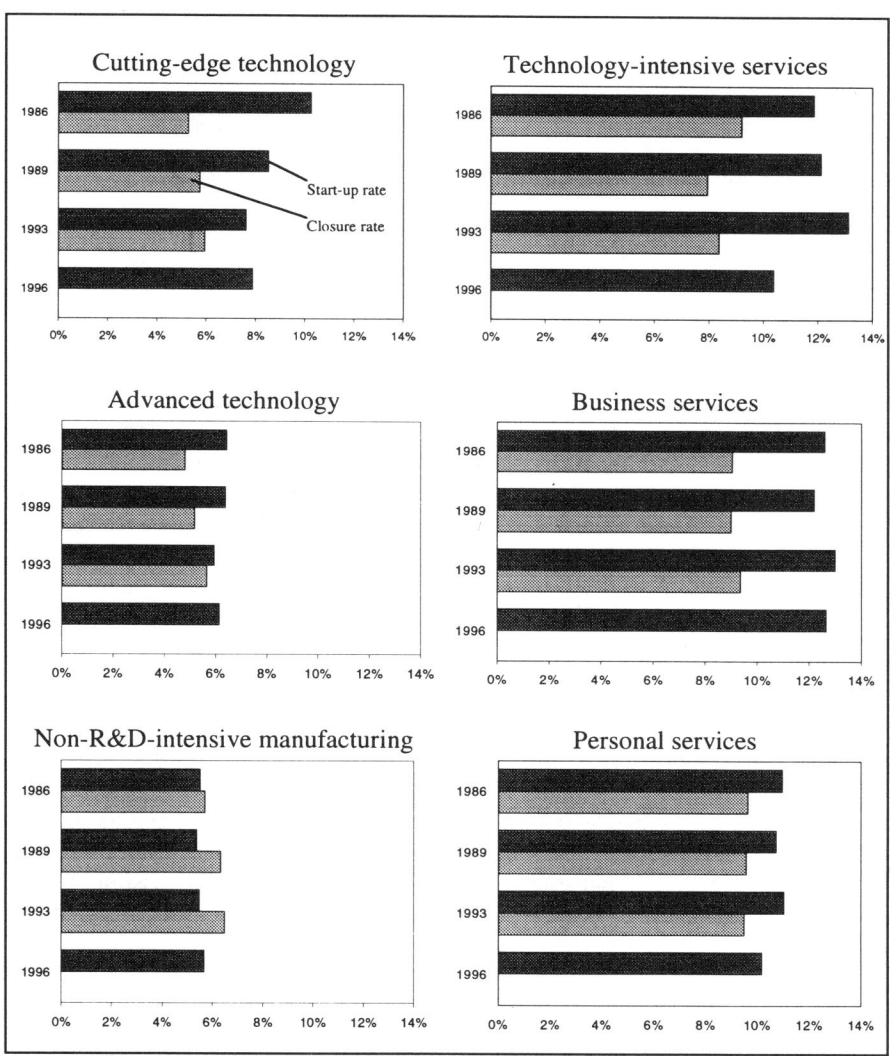

[64]   Figures on the number of company closures in 1996 are not available. The start-up
rate is calculated by dividing the number of new companies founded during a
particular year by the number of companies that already existed on 30 June of the
previous year. News companies are defined as those companies whose identification
number could not be found in the last three years' statistics on employees who are
liable to social security.

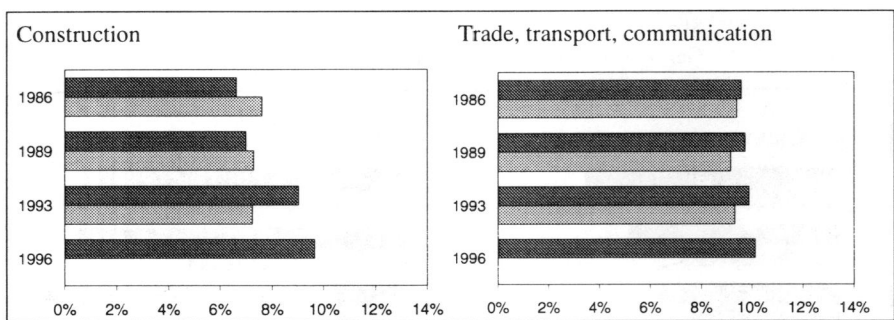

Source: IAB: Statistics on employees who are liable to social security insurance;
          ZEW calculations

The trend toward higher levels of qualification is due primarily to the growing
demand for highly qualified labor. Modern products and production processes
demand increasing amounts of education and know-how in all industries. For this
reason, when jobs are shed, those jobs that require **greater qualification** are
generally not being eliminated as quickly. And in some industries, the demand for
highly qualified personnel has actually increased, while the demand for less
qualified workers has declined. The growing field of know-how-intensive **services**
in particular (Section 4.1.5) demands more highly skilled workers. This is coupled
with new demands on workers' training and their level of vocational qualification.

The trend toward the "service society" is founded on the ever-growing number of
intermediate products from the service sector and on increased service activities
(such as for planning, research, marketing, financing, software, maintenance,
training and logistics) which are in ever greater demand within industry to
augment manufacturing activities. Industrial firms that are based in Germany
largely act as providers of integrated services for all aspects of a particular
product. Germany continues to have a comparatively small service **sector** because
many industrial companies render their own services. Consequently, these services
fall under the category "industrial sector." The level of "service **provision**" for the
general public is also lower than in most other highly developed economies.

Services are becoming increasingly widespread in industrial enterprises; the need
for manufacturing activities is diminishing. Service intensity in the manufacturing
industry averaged approximately 35 percent in 1997 (Table A.7). In all R&D-
intensive industries, the share of employees working in non-manufacturing areas
already exceeds this level, and is in some cases considerably higher. Service
activities have also been markedly expanded in the course of the 1990s in the
relatively production-intensive mechanical engineering, automobile manufacturing
and electrical engineering fields. The intensification of services within industry
goes hand in hand with shifts toward more highly qualified workers (Table A.8).
Most notably, the share of highly trained employees with a college degree has

risen from an average of 15 percent for the entire industrial sector in 1990 (former West Germany) to 19 percent in 1997 (Germany as a whole).[65]

Today, some 72 percent of all employees in the **manufacturing industry** who are **liable to social security** have completed some form of vocational training (Table A.9). As a consequence, the share of low-skilled workers has shrunk by nearly five percentage points during the course of the 1990s. This also reflects the stronger impact that the recession had on this group: Low-skilled workers are more than twice as likely to be unemployed than workers who have completed formal vocational training (see Section 4.3.2.2).

The demand for highly qualified workers in particular is growing (Table A.10). Today, employees with a college or university degree constitute nearly eight percent[66] of all employees in the manufacturing industry, compared to some 5.5 percent in 1990. These figures are considerably higher in individual R&D-intensive industries. The number of highly qualified employees has even risen to 16 percent[67] in the fast-growing area of business, innovation and technology-driven services.[68]

The increase in highly qualified personnel observed in the industrial sector is due mainly to the increased employment of scientists and engineers who have key qualification for technological innovation. However, given that – with the exception of the automobile industry – R&D propensity was very restrained throughout German industry at least until 1996, highly qualified personnel that has been additionally hired has probably been assigned to other advanced service functions rather than to the development of technical innovation.

---

[65]   The increase in the portion of highly qualified employees that developed between 1990 and 1997 when a shift occurred in the areas under consideration (i.e., from former West Germany to present-day Germany) is somewhat exaggerated because of the above-average numbers of college and university graduates among employees in east German industry. However, a comparison of the years 1990 and 1996 in western Germany confirms the trend.

[66]   Differences between the rates of absorption of (highly qualified) college graduates and the data compiled in Section 4.3.2.2 for international comparison are due primarily to the fact that 1) only those employees who are subject to social security and 2) only the manufacturing industry were taken into consideration here. As a result, self-employed persons, freelances, civil servants and government employees were not included in this calculation.

[67]   This rate is even considerably higher in some young, innovative service industries (hardware consultancy services, software and data processing services).

[68]   The growing portion of persons with a higher level of qualification in German industry is also partly due to the overall increase in the level of education among the general population and the larger number of well trained people. With such a labor supply available, companies have workers with various levels of qualification to choose from and often fill positions with highly qualified personnel. This occurs ultimately at the expense of less qualified workers.

An increasing number of college graduates are finding employment in service fields where demand focuses more on scientists and engineers. By comparison, business services, banks/insurance companies and the transport/storage/communication sector are absorbing more and more graduates from other fields.

4.3.2   Education efforts

Highly qualified specialists who undergo continuing training on a regular basis are by far the most important factor determining the success of R&D activities today. They are also indispensable for putting technical know-how to use within a company. The availability of enough (highly) qualified workers lays the foundation for a society's technological performance.

Efforts to quantify education efforts undertaken in Germany or to determine where they fit into international trends are limited primarily to an examination of expenditure on the education process itself. It is assumed here that a positive correlation which is more or less internationally comparable exists between the cost of education and improved qualification. In this connection, when looking at the "output" side, attention is paid primarily to "formal certification" broken down into various rough categories.[69] However, an outsider cannot directly determine whether the national rankings that have been ascertained on the basis of formal criteria also reflect the actual differences in the various countries' "productivity." Furthermore, years may pass between the education system's "input" and the point at which the respective graduate has reached the requisite level of skill and competence, which makes evaluating education system performance on the basis of the indicators at hand nearly impossible.

4.3.2.1   *Germany's education efforts in international comparison*

Expenditure

Germany spent an estimated DM 313 billion in 1997 to expand its knowledge base (this includes first and foremost, government spending on education, continuing education and training, and R&D as well as industry's expenditure on the dual education system, continuing education and training, and R&D[70]). At 8.6 percent

---

[69]   The conceptual approaches to measuring the qualifications of gainfully employed persons use the successful completion of academic and vocational training as a yardstick and additionally take the individual's experience into account. This yields only general information regarding an individual's performance or about the demands of a particular job.

[70]   To be added to this is spending by private households on education and continuing education and training, which can only be estimated at this time.

of the country's gross domestic product, investment in education continues to diminish in Germany (Table 4.5 and Table A.1).

- Public spending on **education and training** in Germany totaled some DM 141 billion in 1997.

- **Industry's** net costs for the **dual education system** amounted to approximately DM 37 billion.

- Corporate expenditure on **continuing education and training** is estimated at DM 10 **billion**. Applying a broader definition – which would encompass, *inter alia*, continued pay during periods of continuing education and training (CET) – industry's CET costs in 1997 can be placed at DM 35 billion.[71]

Government spending on education and training (basic funding for initial training) grew by nearly 70 percent between 1975 and 1990 – more slowly than the gross domestic product (share of GDP in 1975: just under 5%, compared to 3.5% in 1990).[72] Spending on education rose from the start of the 1990s until 1993 due to the immense restructuring requirements in Germany's new *Länder*. The share of GDP represented by education expenditure has however been shrinking since 1993. Government funds for the tertiary education sector have not increased in line with the sector's sizable requirements. Overcrowded classes, adverse conditions, admission restrictions, high drop-out rates and the long time it takes to complete a degree would not seem to indicate that increased efficiency has more than compensated for this lack of funds.

Public and private sector spending on **education and training** (as defined by the OECD, Table 4.6) in Germany represents 5.8 percent of the country's gross domestic product, which is slightly below the average reported by industrialized countries. Yet, when deviations in the age structures of the various countries are taken into account, Germany's spending tends to be above average. This share is slightly lower than in previous years, as is the case in most other OECD countries. Scandinavia and North America spend the most on education – between 6.5 and seven percent of their respective GDP. Measured in terms of per capita expenditure, education is quite intensive in Germany. Nonetheless, Germany cannot afford to allow its education efforts to lag behind those of other highly developed countries competing with it. Otherwise, it is to be feared that

---

[71] This definition was used in last year's report, for example.

[72] Demographic developments play an important role here: Despite a high education intensity (teacher-pupil ratio), lower birth rates have led to less demand at primary and lower secondary level. On the other hand, the number of persons completing secondary education and the number of tertiary level students have increased substantially. The demand for college graduates has grown as well. There has however not been a corresponding increase in funding for the tertiary education sector.

Germany's position could deteriorate if it does not increase its investment in initial and continuing education and training.

International comparison of the structure of investment in education

Germany invests an especially large amount in **secondary level education**, primarily due to industry's sizable investments in the country's dual vocational training system. A system of broad education with high standards has evolved in Germany, rather than the "pyramidal" education structure with its strong concentration in the upper levels to be found in various other countries. The fact that Germany developed this system has to do with to its specific type of innovation capacity which is geared more to the advanced technology field than to cutting-edge technology (see Section 2). On the other hand, the system's focus tends to preserve existing structures.

The USA and other English-speaking countries invest above-average amounts in **tertiary education**. An international comparison of tertiary level education indicates the existence of structural deficits in the German system and in those of all other German-speaking countries insofar as a university degree is very expensive due to the long time it generally takes to complete a degree program. Obtaining a degree in Germany costs nearly twice as much as the OECD average, requiring considerably more than what is usually spent in France, Great Britain or Canada where it takes much less time to complete a degree. In Germany, only *Fachhochschulen* – practice-oriented technical colleges – offer shorter degree programs.

One advantage that shorter time frames offer is that young graduates bring their knowledge and know-how into the innovation process at an earlier stage, which means that the respective economy can utilize and benefit from their knowledge for a longer period. On the other hand, German graduates acquire a higher level of qualification through their longer studies. This in turn translates into a correspondingly higher level of productivity. The issue of "short versus long studies" involves fundamental differences in societal notions of vocational training which compete with one another and have different consequences:

- The USA focuses investment on providing a short, general education which has to be "added on to" later through specific updating training, depending on the individual's vocational needs.

- In contrast, initial education in Germany is long and is very strongly geared to specific occupational demands.

Table 4.5:   Expenditure on R&D and education in the Federal Republic of Germany 1992 – 1997

	In billions of DM					
	1992	1993	1994	1995	1996 [1]	1997 [1]
(1) Research and	76.2	76.6	77.2	79.5	81.0	87.0
Broken down by funding sector:						
Government	27.3	28.0	28.6	29.3	30.1	30.4
Private non-profit organiz.	0.3	0.2	0.3	0.2	0.2	0.2
Foreign	1.5	1.3	1.3	1.5	1.5	1.6
Corporate	47.0	47.1	47.0	48.6	49.2	54.8
Broken down by implementing sector:						
Government; non-profit organiz.	10.8	11.5	11.6	12.3	12.3	12.4
Universities	13.2	13.8	14.4	14.4	15.0	15.1
Corporate	52.3	51.2	51.2	52.8	53.6	59.5
(2) Education and training [2]	162.6	174.0	177.3	188.3	193.0	194.4
(a) Government: [3]	120.7	129.3	131.7	138.3	140.7	140.8
Schools, preschools [3]	85.5	93.2	94.8	99.5	101.3	101.3
Universities [3,4]	15.9	17.0	16.9	18.6	19.2	19.6
Funding for the education syst. [3,5]	7.3	6.7	6.4	6.2	6.0	5.7
Allowances for civil servants [6]	12.0	12.4	13.5	14.0	14.2	14.2
(b) Firms [7]	29.8	30.1	30.6	33.3	35.3	36.6
(c) Private households [8]	12.1	14.6	15.0	16.8	17.0	17.0
(3) Further educ. and training [2,14]	34.4	33.9	31.0	31.7	32.6	31.1
(a) Government: [3,9]	20.1	19.6	16.6	17.3	18.1	16.7
Other education sys. [9,10]	4.0	4.2	4.2	4.3	4.4	4.5
Bundesanstalt für Arbeit [10]	13.8	13.1	10.1	10.7	11.4	9.9
Civil service [11]	2.3	2.3	2.3	2.3	2.3	2.3
(b) Firms [12]	10.2	10.2	10.2	10.2	10.2	10.2
(c) Private households [13]	4.1	4.1	4.2	4.2	4.2	4.2
Total from (1) + (2)	238.9	250.5	254.5	267.8	274.0	281.4
(1) + (2) + (3)	273.3	284.4	285.5	299.5	306.6	312.5

1. Education and training estimated on the basis of the education budget.
2. By sector providing the funding.
3. Basic funds excluding subsidies for civil servants.
4. Excluding R&D conducted by universities.
5. Financial assistance for students.
6. Plus an estimated DM 1.7 billion for allowances paid during times of illness.
7. Industry's net costs for initial training in the dual education system (BMBF projection) plus payments made by industry
   to universities, excluding R&D funds..
8. Revenues of public schools and preschools, and funding for the education system plus payments made to private schools,
   preschools and universities (estimated to be as high as government payments).
9. Adult education centers, libraries, academies.
10. Estimated to be 50% of the Federal Employment Service's total spending on vocational education.
11. Estimated using the average value/employee in the companies (cf. footnote 12).
12. DM 10.2 billion estimated for further training courses in 1992/1993 and 1995 based on a corporate survey conducted by the BIBB and
    Using a broader definition and including sick pay in its calculation, IW arrived at some DM 34 billion for 1995.
13. Estimated to be DM 3.6 billion for 1992 on the basis of GSOEP plus the revenues of public facilities in other education systems.
14. Expenditure solely for education. In other words, excluding sick pay and excluding subsistence payments by the Federal
    Service when individuals undergo further training. When opportunity costs are included, the results of this calculation are much higher.

Sources: DIW compilations, calculations and estimates based on data provided by SV-Wissenschaftsstatistik, the BMBF, Federal Statistical Office, the Bundesinstitut für Berufsbildung (BIBB = Federal Institute for Vocational Training) and the Institut der deutschen Wirtschaft (IW = Institute of the German Economy).

Although an initial degree acquired in Germany is generally to be rated higher than a Master's degree earned in the USA, a German degree does entail disadvantages in respect to occupational, sectoral and regional mobility: Specialized human capital is more difficult to substitute than general human capital, the "mind-to-mind" transfer of technology and know-how is slower, and regional ties grow with age. These disadvantages are particularly serious in times of technology surges when opportunities for applying specific specialized know-how are no longer given. An education system that is geared to narrow occupational specialization is efficient only when vocational demands change but slowly, sectoral structural change can be foreseen on a medium-term basis, and technological change is primarily incremental. Adjustment problems are inevitable in times of rapid structural change. Since job-specific qualification plays a key role in Germany, job switches from one industry to another are a rather seldom occurrence. This situation is different in the USA and France, for example, where the focus is on obtaining general qualification.[73]

Table 4.6:   Education and training indicators in selected OECD countries 1995

	GER	USA	JPN	FRA	ITA	GBR	For information: OECD average[1]
Expenditure on education as a percentage of GDP							
Total	5.8	6.7	4.7	6.3	4.7		5.9
Including:							
Public Expenditure	4.5	5.0	3.6	5.8	4.5	4.6	4.7
Expenditure per student							
in PPP$	5,972	7,905	4,991	5,001	5,157	4,222	5,206
As a percentage of per capita GDP	29	30	23	25	25	24	26
Including:							
Early childhood	21	n. a.	11	16	17	28	16
Primary education	16	20	19	17	24	19	18
Secondary education	n. a.	26	20	31	27	24	25
Tertiary education	43	61	40	33	26	40	49
Tertiary education							
Expenditure per student in PPP$	8,897	16,262	8,768	6,569	5,013	7,225	10,444
Average duration in years[2]	5.1	3.5	-	4.7	4.2	3.4	4.1
Expenditure per degree in thousands of PPP$	45.0	56.9	-	30.8	20.8	24.6	28.9[3]
Highest level of education completed by 25 to 64-year-olds, in percentage							
Early childhood, primary and lower secondary education	19	14	-	40	62	24	40
Upper secondary education	60	52	-	41	30	55	40
Tertiary education	22	·34	-	19	8	22	23

[1]   Including the new member states Mexico, South Korea, Czech Republic, Hungary and Poland.

[2]   1994 (USA 1993).

[3]   Unweighted average.

Sources: OECD: Education at a Glance, Paris 1998; DIW calculations.

---

[73]   Cf. Chr. F. Büchtemann and K. Vogler-Ludwig (1997), Das deutsche Ausbildungsmodell unter Anpassungszwang: Thesen zur Humankapitalbildung in Deutschland.

4.3.2.2   Training levels in international comparison

Training levels

Judging from the level of training among its work force, from the intensity and quality of various types of training it provides, and from its above-average share of scientists and, most particularly, engineers, Germany still comfortably offers the conditions necessary to ensure that it will continue to be among the front-runners in technological performance in the future. However, the number of scientists and engineers among **young** gainfully employed persons is relatively small and could possibly even decline. The first shortages of graduates with degrees in relatively "new" specialties such as I&C technology are beginning to emerge (see also Section 4.3.2.3). Companies also consider such shortages to be a serious impediment to innovation.

Table 4.7:   Highly qualified employees as a percentage of the total work force in selected industrialized countries 1991–1998

		1991	1993	1995	1996	1997	1998	Increase 1991 - 1997
GER (W. Ger.)	Hochschulabschluß	11.7	12.9	13.9	15.1	15.6		3.8
USA	College Degree	27.0	27.7	28.7	29.1	29.4	30.1	2.4
FRA	Diplôme supérieur + Baccaleauréat + 2	16.7	19.0	20.5	21.3	22.0	22.5	5.2
	Diplôme supérieur	8.2	9.3	10.0	10.4	10.9	11.2	2.6
NED	University education + Vocational	21.9		25.3	26.0	26.3		4.4
	University education	6.9		7.9	8.3	8.7		1.7
BEL	Enseignment	8.4	8.7	9.1	9.6	10.3		1.9
GBR	Degree or equivalent + higher education qualification *	20.0	22.0	23.0		24.8		Ca. 5
	Degree or equivalent	12.0	13.0	14.0		15.1		Ca. 3
SWE	Third level of education >= 3 years + postgraduate	11.8	12.1	13.4	13.3	13.7		1.9
ESP	Superiores, Nivel anterior al superior (>= 3 years)		12.5	14.1	15.3	15.9	16.6	
	Superiores (>=5 years)		6.2	7.3	7.9	8.2	8.7	

* Higher qualification includes nurses and elementary school teachers.The percentage of college graduates out of all gainfully employed persons was calculated for FRA (1991-97) and GBR (1991-95) using unemployment rates broken down by level of education and the qualificational structure among the work force.

GBR: Only gainfully employed and/or highly qualified women between the age of 16 and 60. USA: ages 25 - 64; otherwise from 15/16 to 64/65 years of age.

Sources: Former West Germany: Federal Statistical Office, Fachserie 1 Reihe 4.1.2, figures for April; USA: Bureau of Labour Statistics, annual average; FRA: Le marché du travail, second quarter; NED: Statistical Yearbook of the Netherlands; BEL: Statistiques Sociales; GBR: Labour Force Survey; SWE: Statistical Office Sweden; ESP: Boletin de Estadisticas Laborales; ZEW calculations.

International studies pay particular attention to the percentage of college graduates in highly developed economies. At first glance, Germany would seem to exhibit deficits when compared to the USA and other countries because college graduates account for a comparatively low 15.5 percent of its labor force. A closer examination reveals however that shorter courses of study are often offered in other countries, which means that the training provided naturally cannot be as intense as in Germany, thus making a comparison between courses of study available in Germany and those offered in other countries virtually impossible (Table 4.7).[74] This is particularly true of the training offered in Anglo-Saxon countries.

It is striking that France and Belgium report a markedly lower percentage of college graduates in their respective work forces – 11 percent – than Germany, even though their standard period of study is shorter.[75] Irrespective of any problems associated with definitions or comparisons, it should be noted that the percentage of highly qualified workers among the overall work force is growing in all developed economies. Accordingly, the intensification of know-how is a global phenomenon.

Training levels and unemployment

Even though unemployment now affects all qualification levels, investment in education continues to be the best insurance against joblessness: Master tradesmen/technicians and college graduates report five percent unemployment. The rate of unemployment among people who have completed training in an apprenticeable trade is eight percent, undercutting the 18-percent rate registered for people with no vocational qualification by more than half. Nevertheless, there are marked differences between countries: Whereas the unemployment rate steadily falls with higher levels of training in the USA and Great Britain as well as in Belgium and France, it remains relatively high from the master tradesman/technician level onwards in Germany and the Netherlands (Table 4.8).

In Germany, the employment threshold has risen over the years for people without a college degree. In fact, there is no longer a link between growth and employment for those people who have not completed training in an apprenticeable trade (Chart 4.14). Even among people with certification in a recognized trade,

---

74   The availability of shorter courses of study has the effect of, for example, doubling the share of highly qualified workers in France and Spain, and of tripling their share in the Netherlands. Further, the standard period of study varies from country to country: Namely, three to four years in Great Britain, four in France and five in Germany.

75   For this reason, an indicator for measuring a country's endowment with academic know-how that would be suitable for use in international comparisons should be developed by weighting degrees/diplomas with the respective duration of study.

employment has not increased in the last five years despite economic growth. With each year, greater economic growth has been necessary to get an additional number of people into jobs or to prevent unemployment. This applies in particular to persons without vocational qualification. The picture is different in the USA. Due to the particular job market situation there, employment prospects for individuals with an average level of qualification are scarcely any worse than those for college graduates. Employment among low-skilled workers even continues to increase slightly despite the fact that the number of workers in this category has declined continually.

For Germany, the most important thing that needs to be done is to create jobs for low-skilled persons. Fundamentally, this will be possible only when greater flexibility, lower labor costs and a suitable wage policy are given. In addition, innovation could serve as a driving force for realizing indirect "complementarity effects": Innovation and the employment of highly qualified workers could generate new fields of activity which, by increasing the volume, could also offer job opportunities for low-skilled workers.

Still, solving this problem will not be easy. In order to do so, education policy must deal with two challenges: Firstly, it must create new opportunities for people who have seen their employment options reduced as a result of structural change. Secondly, it must develop training opportunities for people who do not meet the requirements of Germany's dual education system. Education is the best insurance against unemployment. However, if structural unemployment among low-skilled persons is to be reduced, consequences will have to be drawn first and foremost in the area of labor costs.

Continuing education and training

Ongoing continuing education is also becoming increasingly important because the half-life of know-how – the period during which know-how can be economically exploited – is becoming ever shorter. It is difficult to evaluate Germany's position in an international comparison of continuing education and training because data on participation in continuing education and training courses vary. Compared to other EU member states, German companies offer a very large number of CET courses. Enrollment rates are however relatively low. All available information indicates that the number of people attending continuing education and training courses has risen substantially since the 1980s.[76] It also indicates a possible trend toward offering courses that are more "made-to-measure" because the ongoing trend toward cutting back on such courses – over

---

[76]   Initial findings from the current "Berichtssystem Weiterbildung" study by the Federal Ministry of Education and Research also indicate that participation in CET measures has increased further since 1994.

half of which are conducted in the individual's own company – continues unabated.[77] The higher the individual's level of qualification (cumulation of education), the more likely he is to undergo continuing education and training: University graduates and persons who have completed non-academic tertiary education have the highest participation rates (nearly half of these people attend at least one continuing vocational training activity a year).

Table 4.8:   International comparison of unemployment broken down by education level

USA	1993	1997	GBR	1995	1997
Less than high school	10.8	8.1	No qualification	14.5	12.0
High school no college	6.3	4.3	Below GCSE grades[a]	10.7	8.1
Some college (1 to 3 years)	5.2	3.3	Further education/GCSE grades	7.3/8.1	4.9/6.8
College graduates 4 years or more	2.9	2.0	Higher education and degree[b]	4.2	3.3
			Post-graduate	n.a.	2.8
GER (former West Germany)	**1993**	**1997**	**NED**	**1993**	**1997**
No vocational certification	12.4	18.1	Primary /Jr. gen.secondary/Pre-vocational educ.	13.4	12.1
Completion of apprenticeable trade	6.2	8.2	Senior vocational college	5.1	4.6
Master tradesman / Technician	3.9	5.2	Vocational colleges	5.0	4.0
Fachhochschule (practice-oriented tech. college)	4.3	4.9	University education	6.0	5.0
University education	4.2	4.9			
BEL	**1993**	**1997**	**FRA**	**1993**	**1997**
Enseig. Primaire ou non	11.8	21.0	aucun diplôme ou non déclaré	18.8	20.9
Enseig. Secondaire supérieur/inférieur	7.3	10.8	CAP ou BEP ou équivalent (vocational college)	10.4	11.0
Enseig. Supérieur non universitaire de type court/long	4.1	4.7	Diplôme de niveau supérieur +		
Enseig. Universitaire	3.3	3.8	baccalauréat et deux ans	6.6	7.9

a GCSE: General certificate of secondary education.

b Higher qualificaton includes nursing and teaching qualification. Data for the year 1993 is not available for Great Britain.

Sources: USA: Bureau of Labour Statistics Data; FRA: Le marché du travail; NED: Statistical Yearbook of the Netherlands; former West Germany: Federal Statistical Office - Fachserie 1 Reihe 4.1.2; BEL: Statistiques Sociales; GBR: Educational Training Statistics for the UK; ZEW calculations.

Funds for continuing education and training should not be diverted from funding for initial education. Rather, they must be allocated additionally. To a large degree, continuing education and training is a kind of "capital spending on replacement" which is necessary for keeping the individual's level of know-how up to date. However, initial training will continue to be a central factor in ensuring that individuals can be educated, are able to practice an occupation, and can generate innovation. The reason for this: Know-how that cannot be sufficiently exploited has to be remedied at some later point through further education. This is costly. On the other hand however, the better initial training is, the more likely the

---

77   In light of the growing importance of continuing education and training, it is of special consequence that expenditure in this area is so difficult to quantify (given that it covers both traditional forms of instruction such as lectures and "soft" forms such as on-the-job learning, etc. which are becoming ever more important in a society that is increasingly based on knowledge and know-how). Accordingly, available calculations reach a very broad range of conclusions.

individual is to undergo continuing education and training, and the more lasting the gains are. It is especially important that workers with low initial qualification levels be induced to attend continuing education and training courses.

Chart 4.14:   Economic growth and employment, by level of education

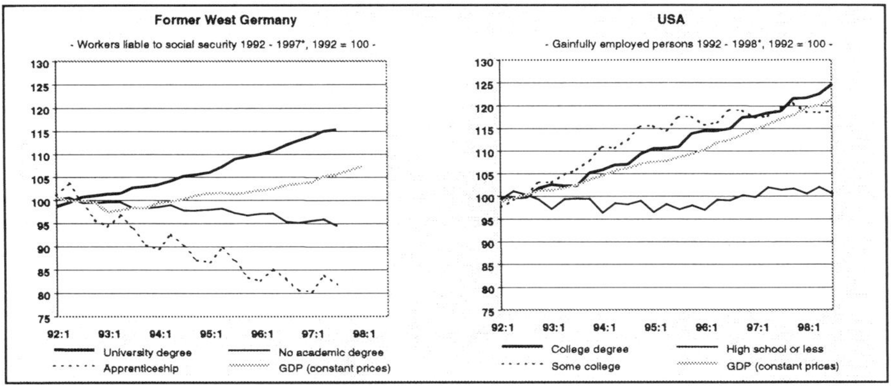

* Average for the quarter.

Sources: Federal Statistical Office; ZEW and NIW calculations; US Bureau of Labour Statistics.

4.3.2.3   Future trends

Education efforts must be evaluated in light of today's growing need for well-trained workers. It is likely that many vocational skills have undergone some devaluation. Manual labor is becoming less and less important in manufacturing, as are simple jobs in the administrative field. In return, activities that entail directing, controlling and regulating are becoming more important. In the service sector, this list is augmented by consultative (and support) activities. R&D activities are also becoming increasingly important in view of the growing pressure to develop innovation. Growth in this area will however largely concentrate on skilled R&D employees. The demand for workers without vocational qualification will probably decline from a current level of 17.5 percent to approximately 11.5 percent by the year 2010.[78] On the other hand, demand for workers who have completed some form of vocational training is growing proportionately. Demand for college and university graduates will see a much larger increase: By the year 2010, 17 percent of the demand for labor (that is

---

[78]   Prognos AG (1997/1998), Arbeitslandschaft der Zukunft - Quantitative Projektion für das IAB.

liable to social security) will target college and university graduates (compared to 10% in 1990 and 7% in 1970).[79]

Consequently, the demands placed on the German education system will continue to grow because demographic trends will require people completing their education in the future to work longer than the current work force will. In other words, there will be fewer opportunities to balance out the supply and demand for higher levels of qualification as one generation succeeds the other because the "generational turnover time frame" will be longer. Rather, greater use will have to be made of intersectoral, occupational and geographical mobility on the part of the work force to bring supply and demand in line with one another. However, the need for greater mobility is at odds with the acquisition of job-specific qualification which companies often desire and which is extensively practiced today.[80] In this respect, it is extremely important to avoid making mistakes in education policy during this phase. We cannot reduce our efforts in the education and training fields, even if Germany's population is aging and fewer children are being born. On the contrary, these efforts must be stepped up so that the growing need for skilled labor can be satisfied. Steps must be taken to counter shortages of highly qualified workers and to prevent unemployment among workers with little training or the wrong kind of training.

Efforts must be made to increase the education system's efficiency – through reform and increased funding – and thereby enhance its productivity. Spending on education, science and research can be considered consumption expenditures from an accountant's point of view at most. From an economic point of view, such expenditure is an investment in growth and employment and, as a result, in raising government revenue levels and reducing the costs it bears from unemployment. Given this situation, it would be even more alarming were expenditure on education to be cut or left to stagnate, and were it to become even more difficult with each passing year to find apprenticeships for all applicants on the traineeship market. Even though supply and demand on the traineeship market may be balanced on paper, the results of this situation are often unsatisfactory for both apprentices and industry. Service fields in particular bemoan the lack of flexibility in adjusting official occupational descriptions and in developing new apprenticeable trades – especially in the area of I&C occupations.

In addition, it is currently not entirely certain that there will be enough graduates in those fields that are important for Germany's technological performance. Germany already has relatively few **young** scientists and engineers and their share will shrink even further in coming years because today's students prefer other

---

[79]   When **all** gainfully employed persons are taken into consideration (this would include freelances, self-employed persons, civil servants, etc.), 17% of the work force today has a college degree.

[80]   Cf. also H.-H. Härtel and R. Jungnickel (1998), Strukturprobleme einer reifen Volkswirtschaft. Analyse des sektoralen Strukturwandels in Deutschland.

fields, as predictions based on Germany's current student population would indicate. When choosing a course of study, students are often guided by their future prospects for employment. This in turn produces relatively strong cyclical fluctuations, particularly in the number of Germany's scientists and engineers.[81] From a student's point of view, the choice of field is an investment decision. The number of first-year students in scientific or technical fields has fallen by nearly half since 1990 (from 40,000 to 25,000 in mathematics and science, and from 75,000 to 40,000 in engineering). Consequently, it is likely that a shortage of persons with qualifications that are essential to the **technical** innovation process could develop within the next few years (Chart 4.15). Shortages are already widespread in I&C occupations today and it will hardly be possible to compensate for them any more.

- Shortages of college graduates also pose long-term problems for company structures. People with a degree in engineering or science in particular are likely candidates for **founding** technology-based companies. In this respect, a skeptical rather than optimistic prognosis is called for here in view of the current level of students' interest in these fields.

- International firms have always considered Germany's large, high-quality reservoir of scientists and engineers to be one of its most important locational advantages. We could potentially see an inversion of this situation when the requisite human capital cannot be obtained by other means.

- In the short term, it will be impossible to avoid shortages at the start of the next decade because students have already made their decisions. However, steps should be taken to prevent universities and colleges from closing down or institutes from merging all too quickly because this would eliminate training capacity on a medium-term basis as well.

- On the other hand, given that a large degree of the inflexibility found today is the result of the length and specialization of tertiary-level studies, efforts should be stepped up to create substitution opportunities between occupations which would make it possible to off-set shortages by making use of post-graduate studies, updating training, continuing education and training and the like. At the same time, steps must be taken to further internationalize education and training and to gear them more toward the standards maintained by those countries that are international leaders in the education field.

---

[81]   The trend is not all that much more stable even in occupations in the health care and education systems where government demand plays a large role.

4.3.3   Science and research

A country's science and research system provides a fundamental platform for its technological performance. The know-how for technological innovation is produced by universities and non-university research centers which train the scientists and researchers who will contribute to the transfer of technology, start up their own companies and put their know-how to use in these firms or, at some later time, in industry and innovative service companies. The **transfer of personnel** is a particularly essential component in the transfer of know-how.[82]

Further, universities and institutes are a vital **source of information** for many companies' innovation activities. The findings produced by scientific and engineering research being conducted at universities and non-university research centers are especially important for "new" research-intensive, knowledge-based fields of technology such as biotechnology/ pharmaceutical products, microelectronics and new materials.

4.3.3.1   Scientific findings

Government research institutes[83] serve a number of functions. Much of the scientific research they conduct is pursued for reasons other than its potential commercial application. Such research is most likely to be reflected in published basic research findings. For this reason, scientific performance is usually measured using the number of **technical publications** in international journals.[84]

At the top of the list, the USA accounts for approximately one third of all published research findings in the natural, engineering and medical science fields. Germany generates more than eight percent of the global volume of publications, putting it approximately on par with Great Britain and Japan (9%). Germany's share of publications – which is considerably understated by this kind of data[85] – has grown slightly. The incentive to offer research findings for discussion in internationally renowned journals has apparently increased despite the tight staffing situation in Germany's universities and research institutes.

---

[82]   Steps to increase occupational mobility would also include measures which foster the flow of personnel between industry and "research managers" working in the scientific community and government R&D facilities, and consequently spur the transfer of know-how on a "mind-to-mind" basis.

[83]   Universities and non-university R&D facilities accounted for just under 32% of Germany's total gross domestic expenditure on R&D in 1997.

[84]   Regarding the methods used here, see the ISI contribution to the 1996 report on Germany's Technological Performance (Grupp et al. 1997).

[85]   The shares calculated for the USA and Great Britain are quite large because the journals evaluated here have a strong focus on life sciences – not one of Germany's strongest suits. Further, the databases used for researching publications draw heavily on English-language journals. As a result, Germany is underrated by the factor 2.

Chart 4.15:   University graduates in Germany

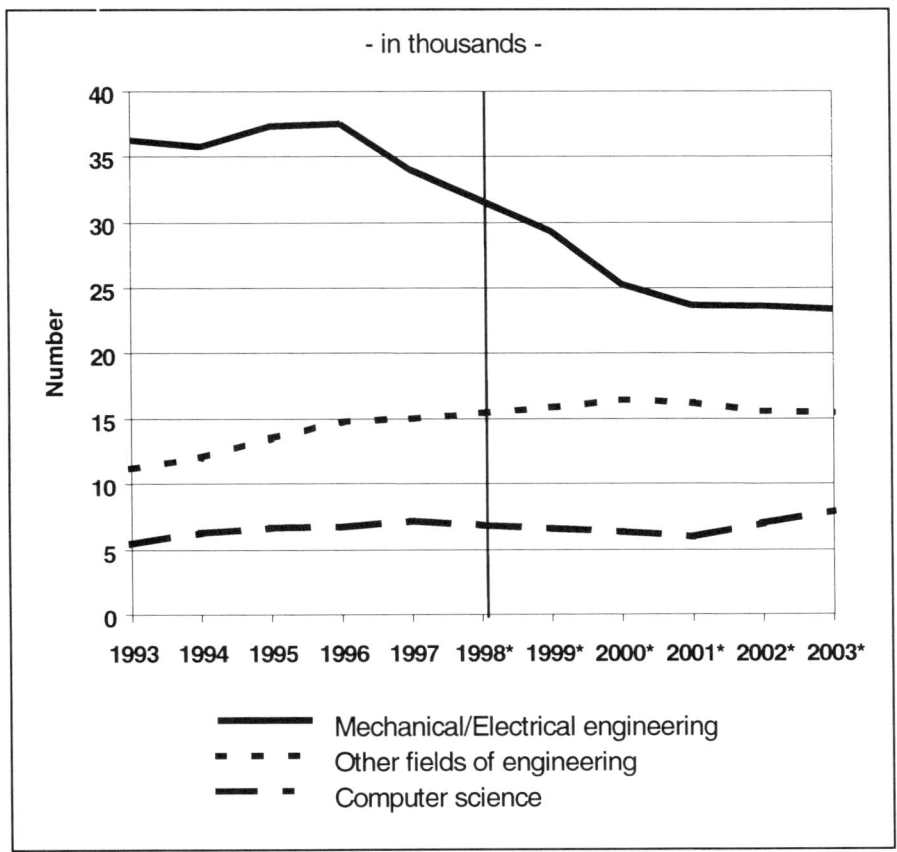

* Prognosis
Sources: Federal Statistical Office; Land Statistical Offices; ZEW calculations

Germany's relative **strengths** are especially well-developed in core areas of physics and chemistry (nuclear physics, solid-state physics/semiconductors, material science, chemistry excluding pharmaceutical science, polymer research and biochemistry). Measurement and control technology and aeronautics/-astrophysics also account for a large number of publications. On the other hand, Germany's strength in electrical engineering, telecommunications and various key fields such as pharmaceutical science/medicine, medical techniques, and information technology falls short of average. By global standards, mechanical engineering/engineering and environmental research in the narrower sense of the term are of somewhat less importance in Germany.[86]

Given that science is becoming increasingly internationalized and that the importance of English-language journals is growing in tandem with this trend, the **international focus** of publications – measured by whether they appear in frequently cited international journals or in journals that receive less international attention – will assume greater weight as a performance indicator in the future (Chart 4.16). For this reason, a primary task will be to increase Germany's presence in internationally renowned journals. By global standards, Germany had reached a slightly above-average level of international focus in the late 1980s. This level fell markedly following German unification in 1991 due to the initially minor exposure of east German scientists in English-language journals. In the years since, east German science has undergone increasing reorientation which has led to a continual improvement in the degree of Germany's international focus.

When the amount of **"attention"**[87] paid to publications is used as a yardstick for measuring quality, German science achieves above-average values – and even places ahead of the USA – despite its structural problems in the tertiary education sector. Particularly striking is Switzerland's very high ranking: It tops the list in terms of both international focus and attention, which is remarkable for a non-anglophone country.

All in all, these indicators allow a positive assessment of the German science system's performance in the area of natural and engineering sciences. Germany's extensive research scene and its reservoir of highly educated scientists are considered to be locational advantages for the country. These two factors are however the result of investment in education and training, manpower and the material outfitting of research facilities during past periods. Given that investment in the scientific field has been on the decline for several years now, only time will tell whether this picture will also be as positive in ten years' time since the low priority assigned to all levels of science has also had a financial impact on basic research.

It must be stressed once again that government-funded research also has an important **teaching and training function**. A consideration of this function should therefore also be included when evaluating research facilities.

---

[86]   Cf. last year's report.

[87]   The level of "attention" is measured by the number of times findings are cited in international journals, adjusted by the respective journal's level of circulation.

4.3.3.2   *Commercial exploitation of research findings*

A number of government-funded research facilities have the explicit task of producing marketable findings and transferring them quickly and directly from the science system to the industrial system. Others are to transfer at least some technology-based know-how. Should a quantification of the direct contribution that scientific facilities make toward technological development be desired, an evaluation of patent applications by universities and non-university research centers would be a suitable **first** step. Such an evaluation must however always take into consideration the special "mission" that the respective institute has to fulfill within the research landscape.

Universities

German universities currently generate four percent of all German patent applications. This is a very high percentage considering that **universities** conduct research not only in technology-oriented fields and that they are firmly geared to basic research (Chart 4.17). The number of patent applications originating in Germany's tertiary education sector has steadily increased over the years and is now four times the number reported in the early 1970s. This development would seem in part to indicate that university research has stepped up its focus on eventual application and its orientation to the industrial sector. This is confirmed by the fact that the amount of funding industry provides for research in the tertiary education sector has seen similar growth. Some 60 percent of the inventions developed in the tertiary education sector are assigned directly to **commercial entities** upon application for patent.[88]

The interlinkage between Germany's universities and its industrial sector presently appears to be better than it is made out to be. Although possibilities for improvement exist in many areas, some fields and institutes have already reached a level of cross-linking where any further increase would jeopardize their mission of conducting long-term research.[89]

---

[88]   In other words, most inventions are being signed over to commercial entities or non-profit institutions before patents are applied for. Patent applications are generally submitted on a **private** basis for those remaining inventions for which a commercial marketing source could not be found.

[89]   Regarding university patenting activities, see the study "Patentwesen an Hochschulen" conducted by G. Becher, Th. Gering, O. Lang, U. Schmoch for the BMBF, 1996.

Chart 4.16:   International focus exhibited by the ten "leading" countries' publications

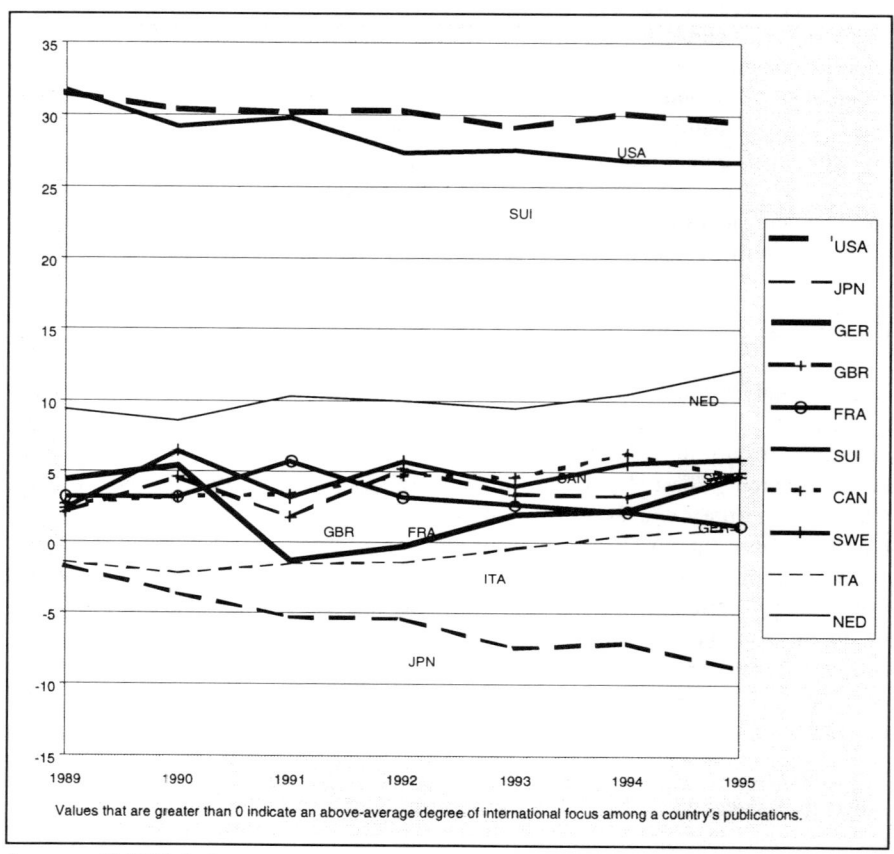

Source: ISSRU (Inform. Science and Scientometrics Research Unit at the Library of the Hungarian Academy of Sciences); FhG-ISI calculations.

Non-university research centers

The number of patent applications being generated by non-university research centers has also risen markedly in recent years. An interpretation of this trend varies depending upon the research facility under consideration. The national research centers belonging to the **Helmholtz Association** cover a broad spectrum that ranges from technically sophisticated basic research to preemptive research all the way to tasks involved in technological development. Today, non-university research centers submit some 360 patent applications a year, twice the number submitted in the early 1990s. Most applications currently come from the Karlsruhe and Jülich Research Centers, the German Aerospace Center and The National

Research Centre for Biotechnology. The focus of patent applications is currently on exploiting "spin offs" – usually chance findings from basic research – for which a suitable partner for exploiting them still has to be found. Particularly since shifting their focus from nuclear energy to other technologies, the Karlsruhe and Jülich Research Centers will first have to develop closer contacts with relevant businesses. Given this situation, patent applications cannot be equated with their conversion into financially successful products. Measured in terms of license agreements, the overall exploitation rates reported by the transfer-oriented Helmholtz Centers would have to be considered marginal. Since these centers have the explicit task of fostering technology within the German research system, the question must be raised: How can potential users be integrated to a greater degree into the research planning phase?

The majority of the patent applications submitted by the **Fraunhofer Gesellschaft,** a leading organization for applied research in advanced technology, develop out of its contract research for industry which corresponds precisely to the Fraunhofer Gesellschaft's mission of transferring technology. Contract research and project funding together represent two thirds of the Gesellschaft's budget. The Fraunhofer Gesellschaft applies for 350 patents every year, four times its rate of submission in the 1980s. Measured in terms of its research budget, the Fraunhofer Gesellschaft applies for a very large number of patents and achieves a very high patent exploitation rate. The Fraunhofer Gesellschaft has assumed a leading role in the transfer of know-how in yet another respect: It gave rise to more than 250 company start-ups during the 1990s.

The **Max Planck Society** pursues a spin-off approach with its patent applications. Although its mission is to conduct top-level basic research, it is still able to report a large number of findings that have industrial relevance in strategically important fields (biotechnology, polymer research, material research). It posts good exploitation rates because it subjects reported inventions to stringent quality control and applies for patents only for those inventions that have relevant market potential. Although it submits fewer patent applications (approximately 50 a year), its exploitation rate is higher than that of the Helmholtz Centers and its license proceeds have risen sharply. Parallel to this, some of the Max Planck Institutes are opening up more to collaboration with industrial partners. Despite this, it is also necessary to review from time to time whether the Max Planck Society's mechanisms for transferring know-how to industry could be improved any.

Germany's "**Blue List**" institutes[90] are strongly geared to science. The Leibniz Institutes are very heterogeneous. As a rule, their missions are not very amenable to the transfer of technology. However, judging by the number of their patent applications, they have become more conscious of applicability. Still, a large

---

[90] With only a few exceptions, these institutes all belong to the Wissenschaftsgesellschaft Gottfried Wilhelm Leibniz science association (WGL).

portion of the 40 to 50 patent applications submitted each year come from the Heinrich Hertz Institute.

Chart 4.17:   Patent applications submitted by universities and other government-funded R&D facilities 1973–1996

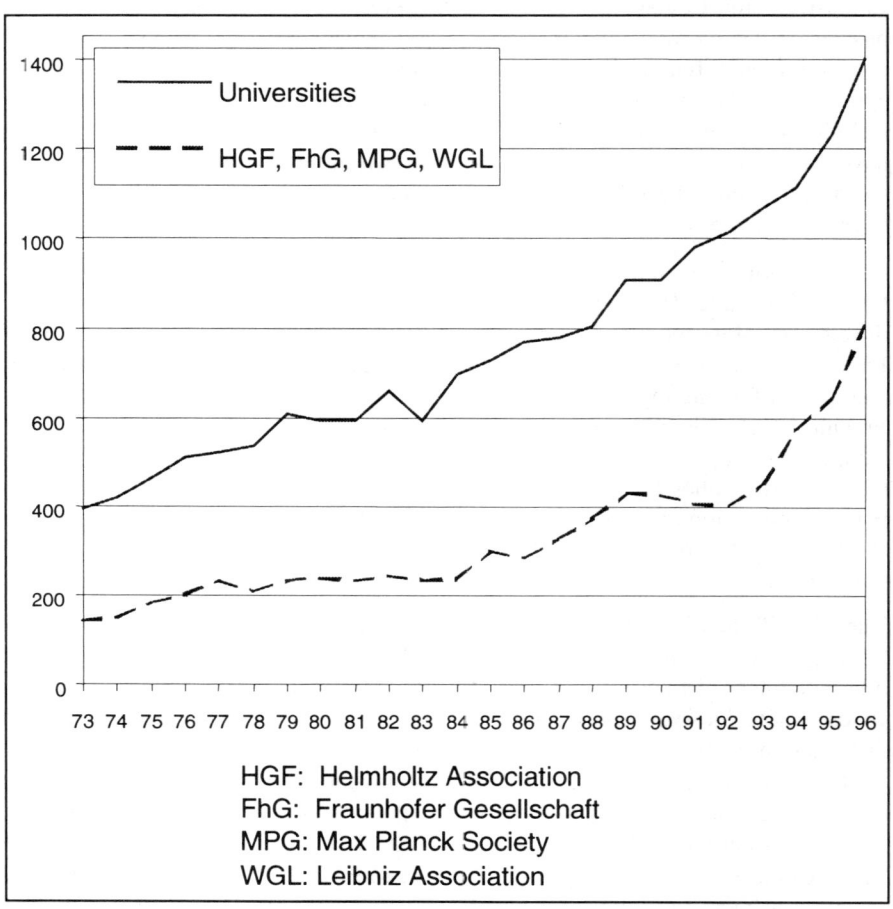

HGF: Helmholtz Association
FhG: Fraunhofer Gesellschaft
MPG: Max Planck Society
WGL: Leibniz Association

Sources: PATDPA; FhG-ISI calculations.

Current challenges

A trend toward making increasing use of patent application statistics to assess application-orientedness and productive capacity among scientific and research centers could be observed in recent years. This entails the danger that government-funded research facilities will also apply for patents for all inventions they report, just to document a supposedly strong focus on application. It is a patent's actual exploitation – as reflected, for example, in license proceeds – rather than the patent application as such that is material to its actual degree of applicability. In view of the enormous cost of the patent procedure, more thought should be given to professional methods for selecting promising inventions and for **systematically evaluating** research findings with an eye to their exploitation.

Corporate restraint in the field of strategic research calls for us to rethink the demands being made of government R&D facilities. It must be remembered however that simply conducting government-funded research is not enough to compensate for a "classical market failure" on the part of industrial research activities. The questions of which direction strategic research should take, which **institutional form** of government-funded research would be effective and what consequences their choices would have for the country's research and science system must be considered. Also to be asked is the question: Which special functions does non-university research fulfill vis-à-vis Germany's universities? This question arises because it would also be possible in a number of cases for large research institutes to hive off areas of basic research that are geared to producing findings. Moreover, scientific research is increasingly geared to approaches that are relevant to potential application. Which is why the traditional, established division of labor between the various centers is subject to gradual change. This division of labor should be subject to a critical discussion and greater competition. This would also entail, for example, having national research centers compete with other research facilities.

The issue of how **interfaces** could be created with industry will be vital because know-how does not automatically transfer itself to industry. The following questions must be answered:

- Do cut-backs in industry's core research activities lead to gaps in know-how?

- Is it at all possible for research being conducted at government research facilities to close such gaps?

- What should the future division of labor between non-university institutes and universities – as well as practice-oriented technical colleges (Fachhochschulen) – look like in the field of application-oriented research?

- Although scientific findings produced by "applied basic research" are generally aimed at long-term exploitation and greater prosperity, would it be possible to translate them into actual applications more quickly?

- Are the cross-links and collaboration between science and industry working well?

- Are we making full use of existing potential? Is the mechanism for transferring know-how via patents, personnel exchange and staff turnover functioning correctly? Are there suitable employment opportunities for scientists and researchers in the industrial sector, particularly in science-based branches?

These questions outline the challenges for the basic funding of the public R&D infrastructure by the federal and *Länder* governments. Which also includes a critical discussion of the relationship between basic funding for institutions, project-related funding and indirect funding. Intensifying **collaboration** between science and industry remains an important task. From a corporate point of view, the transfer of know-how functions well between the public R&D infrastructure and "insider" companies. However, there are still too many "outsiders" who find it extremely difficult to access government-funded R&D facilities that are geared to the transfer of know-how.

4.4   Summary

A survey of current indicators regarding Germany's technological performance reveals the following overall picture:

- Looking at the **short-term**, Germany's innovation system displays a relatively high degree of effectiveness (Section 4.1): New know-how is quickly translated into patented inventions, innovation activity is increasing (on a cyclically neutral basis), Germany's ability to assert itself on global markets is substantial, and the upward trend has stabilized.

- But is this uptrend strong enough? This is where the first shadows fall on an otherwise very bright picture (Section 4.2). Apparently, the **medium-term expectations** on the part of industrial companies are not enough to induce them to significantly expand production capacity in the high-tech sector, regain lost territory in the R&D field or trigger an out-and-out wave of start-ups. This does not mean that some industries are not seeing a lot of activity: For example, biotechnology is doing much to catch up with the international competition, and the automobile industry is busy strengthening its already impressive position. Although the service sector is also sending positive signals in principle, it has nothing outstanding to report on when these signals are compared to international developments. When compared to the USA and other highly developed economies, Germany's service sector still lags considerably behind.

- Even larger problems are beginning to emerge in the **factors that impact technological performance on a long-term basis** (Section 4.3). Although it can be said that Germany's universities and research institutes are highly productive, they are suffering from a "reform jam," with current performance being the product of earlier investment in these areas. At present, not enough is being invested in the future (education, vocational training, colleges and universities). It will be several years until the impact of this is felt. A projection based on the current international situation would not reveal a very positive picture for Germany.

- Inflexibility toward **structural change** also continues to be a problem. Although industrial strengths are being maintained, not enough is being done to mobilize new capabilities in the service sector. Despite gains in the service sector, Germany's stock of innovative and know-how-intensive services is still relatively meager when compared – on a per capita basis, for example – with other countries. Further, the overall service sector is making little progress with growth or jobs. And it has been unable to offset the job cuts undertaken in the industrial sector in recent years. Germany's service sector lacks sufficient internal momentum (see Section 3). As a result, new, diversified and high-powered service markets are not developing adequately (in contrast to the performance being put in by other countries). Which is also why there is not enough demand being placed on manufacturing to generate innovation.

Although industry's outlooks have been positive on a short-term basis, R&D-intensive branches have not been able to reassume the role they played during the 1980s – namely, as a source of new jobs during an economic upswing. The traditional link between growth and employment continues to weaken, even in R&D-intensive industries. The only new jobs being created – when at all – are for qualified persons. Employment opportunities for low-skilled workers have worsened. As a result, qualification patterns continue to shift in favor of qualified persons. Only the service sector is generating new jobs, even though things have been stagnating there for several years. Information and communications technologies belong to the driving forces in this area. However, sectoral mobility among workers in Germany is low by international comparison. Whoever loses his job in the industrial sector today will have a hard time finding a new one in the service sector. Plus, there has been pronounced deterioration in qualification patterns among the unemployed in (western) Germany in recent years. This has not made the scenario for reducing unemployment any simpler.

5 Observations Regarding Special Topics

5.1 Technological competition and Central and Eastern Europe's countries in transition

Newly industrialized countries in today's global markets

Industrialized countries still account for two thirds of global trade today. Despite this fact, centers of global economic growth spread not only within North America in the 1990s, some of them also shifted to Southern Asia during the decade. At the same time, newly industrialized countries were able to substantially improve their positions on international markets for high-tech goods. This group's integration into the global economy has not only increased the pressure on the "simple jobs" factor. The field of players in the global contest over innovation has broadened as well. In particular, competition has gotten tougher in markets with medium innovation potential and larger margins for imitation, where newly industrialized countries can bid from a more favorable income and cost position. On the other hand, highly developed countries not only have to compete on price in these markets but on technology and quality to an ever-growing degree as well. Asia's newly industrialized countries have scored significant success in individual product groups – such as the entire information technology and electrical engineering fields, and photographical and optical products.[91] Expanding beyond their standardized product categories of the past, these countries' export success is due more and more to winning market shares in quality market segments as well.

- As a rule, highly developed countries are also profiting from these shifts in the international division of labor because as newly industrialized countries develop, their import demand for diversified high-quality, sophisticated goods and services grows and is increasingly geared to the range of products offered by industrialized countries.

- Both newly industrialized countries and industrialized countries benefit from the former's integration process because it fosters know-how-intensive growth in industrialized countries.

Last year's report examined the four newly industrialized countries South Korea, Taiwan, Singapore and Israel in detail. Since then, these four countries have – as far as could be observed – further strengthened their R&D performance.[92]

---

[91] Cf. last year's report.

[92] South Korea was still able to expand its R&D expenditure in 1996 – to 2.8% of GDP, compared to 2.7% in 1995. Singapore and Taiwan were also able to make up ground by increasing their R&D intensity from 1.1% in 1995 to nearly 1.4% in 1996 and

However, it is impossible to foresee what effects the current political and economic crises in Southern Asia will have on the fundamental factors that are decisive to technological competition: higher education levels, science, research and development. Given its strong dependence on other countries, this is sufficient grounds for Germany to view developments in Asia with some concern.

The following section focuses on Central and East Europe's five newly industrialized countries Poland, Slovakia, Slovenia, the Czech Republic and Hungary – those former state-trading countries in Europe where the reform process has progressed furthest. These countries are said to have an enormous growth potential which is estimated to be approximately double the growth potential of Western Europe.

Research, development, human capital and patents

Whereas Southern Asia has tended to step up its own **R&D efforts** considerably, Central and Eastern European countries in transition have taken another route to date (Chart 5.1). This region's extremely high R&D intensities fell dramatically with the introduction of free market conditions following the socio-political about-face of the late 1980s. This decline appears however to have come to a standstill since the mid-1990s, at least in Poland, the Czech Republic and Slovakia. The Czech Republic and Slovakia were already conducting R&D at a relatively high level before the peaceful revolutions of the late 1980s, even though their output could not compete with that of western industrialized countries, as was also the case with all other Central and East European countries in transition.

R&D spending at university level was cut by more than half in 1996 in Slovenia, the most R&D-intense country in this group. Slovenia's advanced ranking among newly industrialized countries is evidenced by the comparatively large share that R&D scientists and engineers represent out of its total working population.

R&D's most important source of funding – government – has withdrawn from the scene. Corporate involvement in R&D is strongest in the Czech Republic and Slovakia (Chart 5.1). Only in Poland is the government's portion of R&D funding still quite high, despite the fact that other funding sources (including foreign sources) have expanded their share of total funding. Foreign sources of financing are comparatively important in Hungary in particular where several multinational firms have set up R&D facilities either by making use of existing research capacity or by building new development centers of their own.

The individual newly industrialized countries concentrate their R&D efforts on different sectors. Poland and the Czech Republic tend to give priority to advanced

---

1.8% to 1.9% respectively. At 2.2%, Israel was the only country in this group to report a (slight) decline in R&D intensity for 1996.

technology (mechanical engineering, automotive engineering), whereas R&D efforts in Slovenia and especially Hungary focus on the pharmaceutical industry.

Chart 5.1:   R&D intensity in selected newly industrialized countries and in Germany 1991–1996

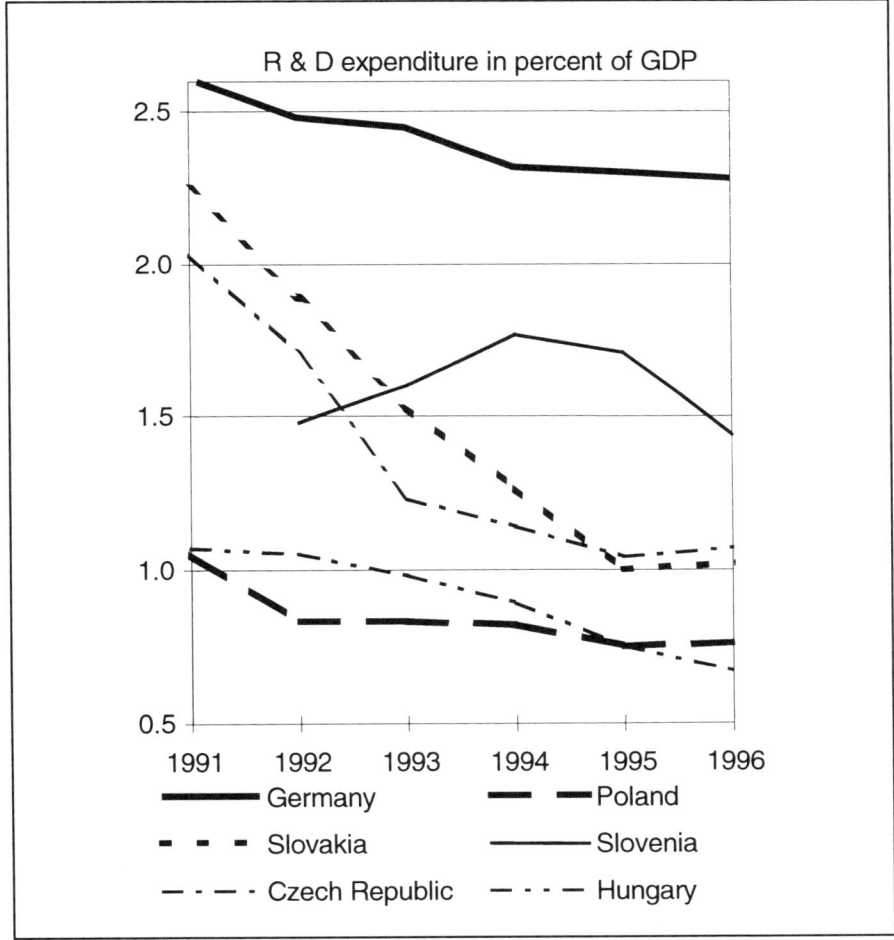

Sources: Local sources; OECD: Main Science And Technology Indicaroes; NIW estimates

Human resources can generally be deemed quite promising – albeit only when viewed in terms of formal education levels. Using this yardstick, education levels in Poland are quite high. This is however relativized by the fact that the country does poorly in studies on the productivity and efficiency of education and training systems in respect to industry needs and in studies on the availability of qualified

148   Measurement and Interpretation of Germany's Technological Performance

Table 5.1:   Indicators on R&D in selected newly industrialized countries of Central and Eastern Europe

	Poland	Slovakia	Slovenia	Czech Republic	Hungary
R&D intensity					
Total R&D expenditure as percentage of GDP 1996	0.76	1.02	1.44	1.07	0.67
R&D expenditure by industrial enterprises as percentage of GDP 1996	0.31	0.57	0.71	0.64	0.29
R&D funding					
Percentage attributable to					
Government	57.8	39.5	43.4	35.3	50.0
Universities/ Private non-profit organizations	1.9	0.1 (w/o univ.)	4.9	2.9	6.5 [1]
Corporate sector	38.9	57.4	49.0	59.6	38.9
Foreign sources ..... in 1996	1.4	3.0	2.7	1.9	4.6
Performance of R&D activities					
Performance of R&D expend. attributable to:					
Universities/ Private non-profit organizations	31.2	39.4	27.7	31.2	32.0
Universities	27.8	5.1	21.6	8.9	24.8
Corporate sector ..... in 1996	40.9	55.5	50.7	59.9	43.2
R&D personnel					
Percentage of R&D scientists / engineers out of total work force	0.31 (1996)	0.39 (1995)	0.52 (1995)	0.25 (1996)	0.26 (1996)
Focus of R&D activities, by sector					
Share of manufacturing companies' R&D expenditure attributable to the respective sector	Mech. engineering 19.5 (1995)	Not available Data initally not made public	Pharmaceutical industry 31.4 (1996)	Automobile industry 32.8 (1995)	Pharmaceutic. Industry 50.5 (1995)
	Chem. w/o pharma 11.7		Elec.Mach'ery / apparatus 18.9	Mechanical engineering 14.2	Pet.products, nuclear fuel 10.4
	Electrical machinery/apparatus 9.4		Radio/TV/ Comm Equip. 17.8	Air- and Space-craft 10.4	Chemicals (w/o pharma) 10.4
Indicators on the deployment of highly qualified employees					
Percentage of university graduates out of total work force in 1996	10.3*	11.3 (1995)	6.8	10.4	6.1
Percentage of senior civilian servants, professional and managerial staff, scientists and technicians[2] out of total work force in 1996	26.7	32.9 (1995)	30.4	34.0	29.4

Errors due to rounding-off.

[1] Other funding sources; these accounted for only 0.7 percent one year earlier (1994: 3,3%)

[2] ISCO-88 (International Standard Classification of Occupations) Primary groups 1, 2 and 3

* Slightly overestimated.

Sources: Local sources; ILO; IMD; OECD; UNESCO; Federal Statistical Office; NIW calculations and estimates.

engineers.[93] Hungary on the other hand posts good results in this regard. Measured in percentage of GDP, government spending on education and training in Poland exceeded the OECD average in 1994; Poland even took top honors in an international comparison of the amount spent per student in the tertiary education sector.[94]

Chart 5.2:   Patent applications submitted to the EPO by newly industrialized countries of Central and Eastern Europe 1985–1996

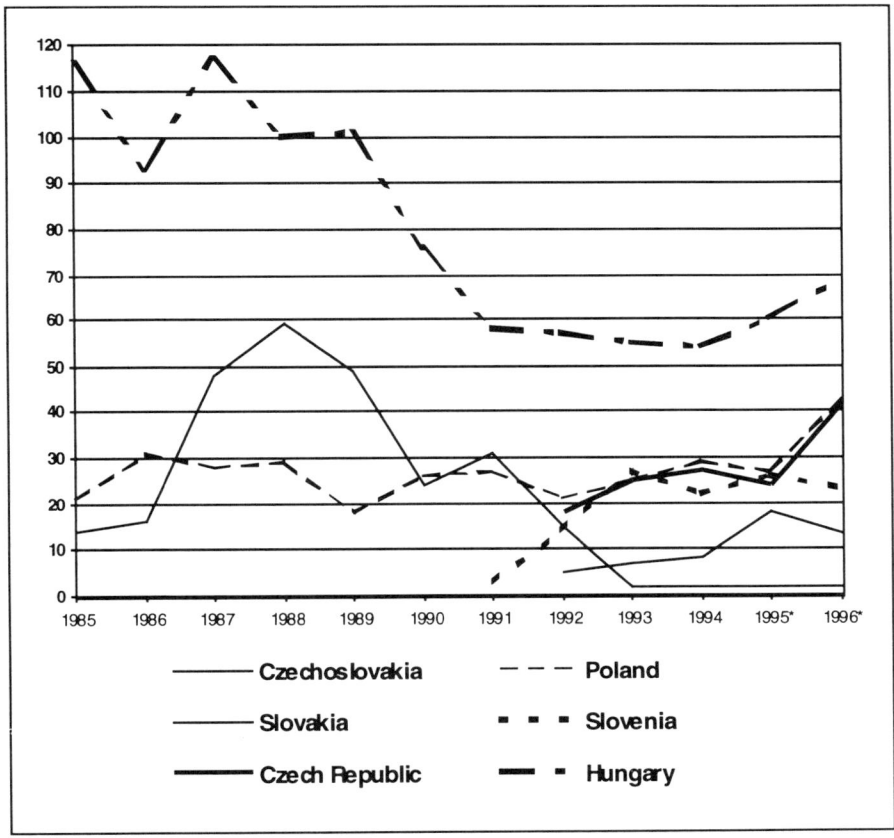

* Projected

Sources: EPAT, PCTPAT; FhG-ISI calculations

Compared with highly developed industrialized countries, general education levels in the Czech Republic are very high, particularly among young people. The

---

[93]   Cf. International Institute for Management Development (1998), The World Competitiveness Yearbook 1998.

[94]   Cf. OECD (1998), Human Capital Investment - An International Comparison.

percentage of gainfully employed persons working in highly qualified occupations is very high in the Czech Republic. Slovakia can be expected to exhibit patterns similar to those found in the Czech Republic. The human capital of this group of newly industrialized countries is likely to provide a solid foundation for expanding R&D activities in the future.

In terms of quantity, the number of **patent applications** being submitted by Central and East European countries in transition approximates the number generated by Asia's newly industrialized countries in the mid-1980s. This number is growing, evidence of the region's increasing innovation capability (Chart 5.2): The level of EPO patent activity on the part of Central and East European countries in transition approximately doubled during the period from 1991/1992 through 1996. Hungary's patent activity level grew the fastest, albeit without being able to reach the levels reported by Asia's newly industrialized countries, such as South Korea and Israel in particular. Although Hungary had already reached a relatively high application level in the 1980s, it has reported the least growth in patent applications since the early 1990s. Between 1992 and 1996, Slovenia and Slovakia however saw their patent activity take up the pace previously set by South Korea, Israel and Singapore.

Foreign trade with high-tech goods

Central and East European countries in transition are presently only somewhat integrated into the global economy. These five countries do not play a role as a target region for the R&D-intensive **exports of industrialized countries**. Germany was however already selling five percent of its R&D-intensive exports to this group of countries in 1996 (Table 5.2). These countries' capital requirements for economic development and environmental rehabilitation in particular fit in well with the range of goods and services offered by German industry, whose export supply pattern closely matches the newly industrialized countries' import demand patterns for these goods, particularly in advanced technologies (Chart 5.3). Only Hungary's import demand pattern deviates from this picture. Compared to Japan or especially the USA, it is easier for Germany to serve Central and Eastern Europe's import markets for high-tech goods not only because of its strengths arising from its locational advantages and its so-called locational proximity advantages, but also because of its often much more compatible range of export products. The current crisis in Southeast Asia illustrates just how important broad regional diversification is. In this respect, it has been to Germany's advantage that R&D-intensive goods have gained so much importance in its trade with Central and East European countries in transition. Nearly 40 percent of the high-tech imports these countries have bought from industrialized countries are of German origin. Geographical proximity, common cultural ground and a revival of traditional trade relations have all been factors in stimulating trade. The countries of Central and Eastern Europe have become interesting as partners for trade and as locations for investment. German trade with these countries is expanding rapidly (and quite evenly) across the entire board.

Table 5.2:   Foreign trade links between the newly industrialized countries of Central and Eastern Europe,[1] non-OECD countries as a group and individual industrialized countries/groups of industrialized countries for trade in R&D-intensive goods – 1996 – in percent

Country	Germany				USA				Japan				OECD			
Product group	Non-OECD countries (of exports)	CEE countr. (of exports)	Non-OECD countries (of imports)	CEE countr. (of imports)	Non-OECD countries (of exports)	CEE countr. (of exports)	Non-OECD countries (of imports)	CEE countr. (of imports)	Non-OECD countries (of exports)	CEE countr. (of exports)	Non-OECD countries (of imports)	CEE countr. (of imports)	Non-OECD countries (of exports)	CEE countr. (of exports)	Non-OECD countries (of imports)	CEE countr. (of imports)
Total manufactured goods	20.0	6.0	16.5	6.5	29.3	0.3	31.3	0.3	44.8	0.2	42.4	0.2	24.5	2.2	20.4	1.7
Cutting-edge technology	19.9	4.3	16.4	1.8	34.4	0.3	36.5	0.3	41.2	0.2	29.7	0.1	29.4	1.4	21.8	0.6
Advanced technology	20.6	5.3	9.8	5.0	26.7	0.3	15.9	0.1	38.1	0.3	29.0	0.1	23.1	2.4	10.5	1.3
Total R&D-intensive goods	20.4	5.0	12.4	3.7	31.0	0.3	24.4	0.2	39.3	0.2	29.4	0.1	25.6	2.0	15.1	1.0
Biotechnology/Biotech. substit. sector*	21.9	4.6	8.6	1.6	24.7	0.3	17.2	0.5	34.6	0.4	13.5	0.5	21.9	2.4	9.0	0.8
Radioactive materials	3.8	10.9	18.9	0.1	5.5	0.0	24.5	0.0	25.2	0.1	7.3	0.0	12.7	0.7	27.1	0.0
Other chemicals	21.6	4.7	5.6	1.6	25.1	0.2	12.0	0.2	45.3	0.1	17.6	0.2	24.0	2.5	6.7	0.9
Mechanical engineering	32.3	6.0	6.6	7.3	35.1	0.5	9.3	0.6	50.2	0.2	19.3	0.1	34.2	3.0	5.9	1.5
Automobiles	10.4	3.2	1.4	3.1	20.6	0.2	0.0	0.0	28.7	0.5	0.1	0.0	14.0	1.9	0.8	1.2
Aircraft and spacecraft	11.3	0.4	5.7	0.4	37.8	0.2	7.9	0.1	16.2	0.0	0.3	0.0	36.8	0.3	12.6	0.3
Information technology	21.0	5.3	24.8	1.7	34.7	0.4	45.8	0.1	38.2	0.2	47.0	0.0	27.8	1.3	31.3	0.5
Electrical engineering, n.e.c.	19.9	8.4	16.4	10.6	22.3	0.4	23.2	0.7	46.8	0.2	35.8	0.2	24.5	2.8	13.9	2.5
Measuring & control technology	19.7	5.6	10.6	4.2	23.5	0.4	17.1	0.1	36.2	0.2	10.3	0.1	22.7	2.1	8.6	0.7
Photographical and optical products	13.8	3.3	20.1	2.6	23.4	0.2	31.2	0.2	32.2	0.0	50.7	0.1	23.2	1.0	23.3	0.6
Other R&D-intensive goods	21.0	6.3	19.9	9.3	52.0	0.1	25.2	0.4	47.1	0.1	31.9	0.1	35.2	2.0	18.5	2.2

[1] Poland, Slovakia, Slovenia, Czech Republic, Hungary.

* Includes bioengineered chemical products and traditional chemical products for which the biotechnology field could provide substitutes.

Sources: OECD: Foreign Trade By Commodities, CD-ROM; NIW calculations.

Chart 5.3:   Level of conformity* between the export structures of selected OECD countries and the import demand of the newly industrialized countries of Central and Eastern Europe in R&D-intensive goods   1996

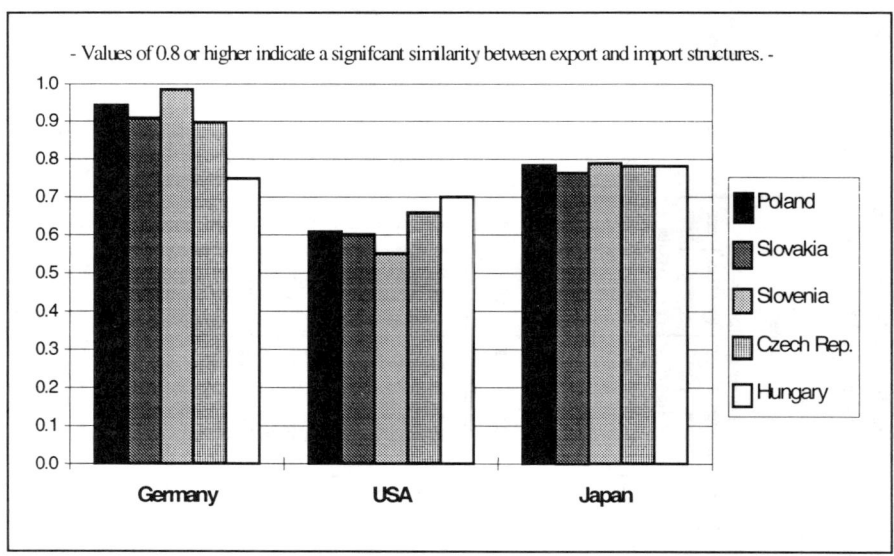

- Values of 0.8 or higher indicate a signifcant similarity between export and import structures. -

Poland

Slovakia

Slovenia

Czech Rep.

Hungary

Germany   USA   Japan

* Preference pattern test: Values correspond to the cosine of the angle between the export and import vectors. The imports of these advanced threshold nations were estimated on the basis of OECD exports to these countries.

Sources: OECD: Foreign Trade by Commodities, CD-ROM; NIW calculations and estimates.

The newly industrialized countries of Central and Eastern Europe have been virtually non-existent on the industrialized countries' **import markets** for R&D-intensive goods to date, accounting for only one percent of this trade. Measured by this yardstick however, Germany is an exceptionally important source of demand for such goods from this region: Some 3.5 percent of the R&D-intensive goods that Germany imported in 1996 came from these countries. Central and Eastern Europe's newly industrialized countries were in some cases able to claim considerable market shares for advanced technology goods in particular. They have been able to win an increasing amount of ground in the advanced-technology goods field, are finding buyers on the high-powered markets of highly developed industrialized countries and are edging some of the goods those countries produce out of the marketplace. Central and East European countries in transition account for more than 25 percent of the electricity distribution and control apparatus and locomotives and rolling stock that Germany imports. Germany's trade with the countries of Central and Eastern Europe is already very intra-industrial in character and offers Germany's R&D-intensive industries further opportunities to specialize.

Chart 5.4:   Price structure of Germany's cutting-edge-technology imports 1996

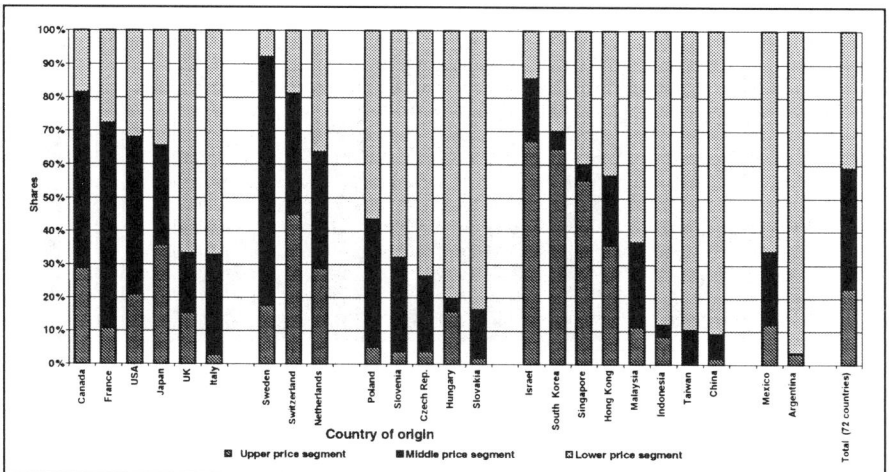

Sources: OECD: International trade by commodities statistics, HS 1988-1996 on CD-ROM; DIW calculations.

There are of course differences between products from western industrialized countries and products from newly industrialized countries. Looking at R&D-intensive goods, for example, a very large portion (up to 50 to 75%) of the R&D-intensive goods being imported from Central and East European countries in transition are to be found in the lower price segment;[95] a maximum of 11 percent of the imports from these countries belong to the upper price segment (Chart 5.4 and Chart 5.5). By comparison, (industrialized) countries with their relatively large reservoirs of well-trained workers produce first and foremostly goods in the upper and middle price segments. Consequently, it must be assumed that the goods being imported from Central and East European countries in transition tend to be at the end of their product life cycle (products such as inexpensive consumer goods whose technology is comparatively unsophisticated), whereas industrialized countries are supplying expensive technology-intensive products (industrial and capital goods).

---

[95]   When the relative import-unit value of a country's exports exceeds the average for the respective product group by 25% or more, they are assigned to the upper price segment. When their relative import-unit value is less than 80% of the average value, these products belong to the lower price segment. All other exports are classified as belonging to the middle price segment.

Chart 5.5:   Price structure of Germany's advanced-technology imports 1996

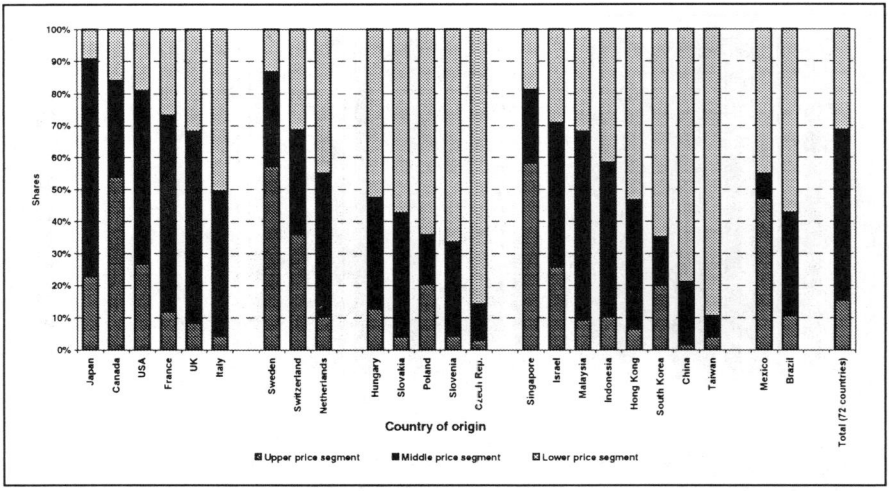

Sources: OECD: International trade by commodities statistics, HS 1988-1996 on CD-ROM; DIW calculations.

5.2   The catching-up process in Germany's new *Länder*

The development of R&D-intensive industries

Germany's new *Länder* are slowly making progress with their integration into international competition over innovation, but are – all in all – still lagging considerably behind the progress reported by the former West Germany.

Exports accounted for nearly half the sales revenue Germany generated with R&D-intensive goods in 1997. By comparison, foreign sales were responsible for one quarter of the new *Länder*'s sales revenue from such goods (Table 5.3). Despite this situation, the R&D-intensive sector in eastern Germany has recently reported remarkable **export success**. The share of sales generated abroad by some east German industries is currently about as large as the share reported by the rest of the country (sections of the chemical industry, photographical and optical products, aircraft and spacecraft, EDP, electronic components, communications engineering). Foreign sales are still however subject to sizable fluctuations in many industries. The fact that small businesses represent the bulk of companies in eastern Germany and that the cash flows of east German firms are generally weak precludes more complex forms of internationalization such as direct investment abroad or joint production activities with foreign partners. Even cross-border commission processing with nearby partners in Poland or the Czech Republic still plays a comparatively minor role.[96] East German firms consider the establishment of stable relations with western firms to be an important route toward market integration.

Table 5.3:   Standing and development of R&D-intensive industries in Germany's new Länder

	Net Output		Employment			Export rate	Foreign sales		
	Average annual change in % 1993-1997		% of D 1997	Average annual change in % 1995-1997		in % 1997	% of D 1997	Average annual change in % 1995-1997	
	New Länder	Former WG	New Länder		Former WG	New Länder	New Länder		Former WG
Total R&D-intensive industry	5.0	2.9	6.9	-6.6	-3.0	25.0	2.5	25.7	10.6
Cutting-edge technology	7.3	3.7	5.6	2.7	-3.9	35.0	3.5	80.0	16.2
Advanced technology	4.6	2.8	7.2	-7.8	-2.8	22.8	2.3	16.1	9.4
Non-R&D-intensive industry	10.1	0.6	10.1	-1.3	-3.7	11.3	4.4	14.0	4.8
Total manufacturing industry	8.3	3.7	8.7	-3.3	-3.4	15.7	3.2	19.6	8.5

* Without nuclear fuel.

Sources: Federal Statistical Office: manufacturing sector statistics; NIW calculations.

In light of the region's industrial structural deficits (small percentage of high-tech industries, company size structure) and the instability still prevalent among many companies, particularly in the capital goods industry, it is not at all surprising to find a less-than-average level of integration into the **globalization process** among east German firms. These deficits can be eliminated only on a long-term basis.

Nonetheless, the increased activity observed in east German export trade is a positive sign. However, in view of the international sector's small size, it is not very likely that this increased activity will lead to the creation of additional jobs. East German operating units accounted for only 2.5 percent (or a total of DM 10 billion) of Germany's total foreign sales by R&D-intensive industries in 1997. Although this is half a percentage point more than in 1996, it still indicates that the region's export base will remain narrow for some time to come.

The growing technology-orientedness of east German industry is evidenced by the **investments** effected between 1993 and 1995 when most companies in the region replaced almost their entire their equipment and vehicle fleets. Although overall investment activity in eastern German has since waned, some research-intensive industries have posted further investment growth (electrical engineering, automobiles, and, somewhat more moderately, chemicals). According to ifo Institute surveys, planning for 1998 indicates that investment will be increased in the mechanical engineering, precision tools/optical products and aeronautics

---

[96]   Cf. in this regard DIW/IfW/IWH (1998), Gesamtwirtschaftliche und unternehmerische Anpassungsfortschritte in Ostdeutschland.

fields, whereas the propensity to invest is likely to be generally somewhat more subdued in less R&D-intensive industries.

The high-tech sector in the new *Länder* accounts for approximately one third of the region's **industrial production**, making it considerably less important than it is in the western *Länder*. Its importance diminished during the course of the 1990s. That portion of industrial production attributable to manufacturers of construction-related products, less technology-intensive goods and difficult-to-market products is distinctly larger and has even grown. At present, it is not yet possible to speak of the R&D-intensive sector as being a driving force in eastern Germany. Positive spillover effects are most likely to come from the mechanical engineering, plastics, EDP, electronic components, and aeronautics sectors. The list of losers includes in particular railroad vehicles, parts of the mechanical engineering industry, medical technology and medicines. However, previous sources for growth in non-R&D-intensive branches in eastern Germany have run dry. The larger companies in the capital goods industry have also found new owners and will be able to serve as a driving force now that their restructuring has been completed.

With their slow growth and strong competitive pressure, R&D-intensive industries in eastern Germany continue to shed jobs at a markedly faster pace than seen in the rest of the industrial sector. Compared to western Germany, the "decoupling" of growth in value added from employment in the R&D-intensive sector has progressed much further in the eastern part of the country, and has led first and foremost to enormous job cuts in **advanced technology** fields. However, a number of large branches which have been enjoying favorable economic conditions – such as the automobile industry, parts of the electrical engineering industry (electrical equipment for motor vehicles, lighting equipment/electric lamps) and tool manufacturing – were able to report staffing increases, at least for 1997. This picture continues to be contrasted by extensive job-shedding in structurally important industries such as locomotives and rolling stock, as well as the metal-working industry and machine tool manufacturing. The service sector in the new *Länder* has yet to generate any impetus worth mentioning for employment because those of its areas which have enormous expansion potential – such as business services – are conspicuously underrepresented in the region. The number of people working in the **cutting-edge technology** field grew in 1996/1997, albeit along a low baseline. This growth is the result of staffing increases undertaken at that time in EDP production facilities and in the aeronautics industry.

Research, development and innovation

Following major cut-backs in the early 1990s, the **core R&D work force** in the new *Länder* has settled in at a level of some 23,000. As a result, the new *Länder* employ something more than eight percent of Germany's total R&D work force while accounting for only some five percent of the country's total R&D

expenditure. This is due primarily to extensive federal and *Land* programs aimed at maintaining well-trained staff in the region's firms: Some two thirds of all east German companies are participating in these programs. It is however to be feared that pending reforms (i.e., the introduction of an SME clause and an ownership clause) will have a negative impact on the number of people employed in R&D in eastern Germany because many of the companies conducting R&D in the new *Länder* do not yet have a very strong equity position and are therefore scarcely in a position to maintain their level of R&D activity without government assistance. These reforms additionally pose a threat to the success of innovation activity because government-assisted firms in particular are much more likely to be more involved in the innovation process. These companies' innovation and market performance is in part better than that of innovative firms which receive no assistance.

Although the number of **firms conducting R&D** continues to be comparatively small, it has grown in recent years and R&D spending is likely to have increased slightly in 1997.[97] The share of external R&D expenditure has grown, indicating that firms are increasingly seeking **R&D collaboration**, particularly in the area of process innovation. R&D activities have also been stepped up in companies that do not belong to a west German corporate group. A breakdown by company size reveals that considerable differences still exist between eastern and western Germany: In former West Germany, large corporations account for 85 percent of internal R&D spending. In the new *Länder*, this figure is only 35 percent. A large portion of the product innovators in eastern Germany use R&D to develop imitations of other manufacturers' products – a perfectly sensible strategy for making headway in the catching-up process. The percentage of companies in the new *Länder* which have been able to cut costs with the help of process innovation has grown at an above-average pace, as has the share of cost savings generated in this way. This is also a reflection of the growing cost-push pressure which these firms in particular have to deal with as a result of the growing globalization of their markets.

The mechanical engineering industry employed a good 20 percent of the R&D personnel working in east German industry in 1997. Another 13 percent were employed in the measuring and control engineering/optical products field, 12 percent in the chemical industry and nine percent in the automobile industry. At 13 percent, the "research and development for other firms" field also accounted for a large portion of the R&D personnel working in the new *Länder*. This

---

97   Both the WSV and the Forschungsagentur Berlin GmbH research agency (C. Hermann, T. Konzack and P. Ständert (1998), Analyse zur Entwicklung der Potentiale in Forschung und Entwicklung im Wirtschaftssektor in den neuen Bundesländern im Zeitraum 1990 bis 1997) come to this conclusion on the basis of differently structured polls they have conducted. In contrast, the ZEW's 1997 innovation survey was somewhat more skeptical.

category covers external industrial research facilities (see below) and innovative firms whose R&D services make up a large share of their total output.

Apart from its structural deficits (percentage of R&D-intensive industries, company size patterns) and the relatively high R&D intensity of its small businesses in comparison to western Germany, it is also typical of eastern Germany that large portions of the actual **research** elements of R&D are to be found in external industrial research facilities. As a result, these R&D resources are not directly available within east German firms, even though these companies can in principle partake in the R&D service providers' findings. Given that large-scale industrial research is not being conducted in eastern Germany, external R&D capacities could play an important role in interlinking innovation activities between industrial research and production. To accomplish this, these facilities must integrate themselves into the international division of labor and find their respective markets as if they were manufacturing companies, something that they have not completely succeeded in doing in the past.

Calculated on a per capita basis, Germany new *Länder* generate approximately one third of the country's total **patent** applications (four applications for every 10,000 employed persons compared to 13 on national average). The growth rates reported for patent applications are higher than the national average throughout nearly all of eastern Germany. Between 1995 and 1997 alone, the number of patent applications coming from eastern Germany grew by nearly 25 percent (Table 5.4). More than half of all firms in eastern Germany that conduct R&D have applied for patents on their R&D findings since 1990.[98]

All in all, **start-up activity** in the new *Länder* is presently comparatively slow. The number of businesses being newly established is even diminishing. This is also the case in R&D-intensive industrial fields (-2.8% in 1997), albeit not at the level reported in the rest of industry (-6.8%). The number of new start-ups in the high-tech service area and in other business services is growing at rates (6.7% and 5.1% respectively in 1997) similar to those reported in western Germany.

---

[98]   C. Hermann, T. Konzack and P. Ständert (1998), Analyse zur Entwicklung der Potentiale in Forschung und Entwicklung im Wirtschaftssektor in den neuen Bundesländern im Zeitraum 1990 bis 1997.

Table 5.4:   Comparison of R&D and innovation ratios for eastern and western Germany

	Year	New *Länder*	Former West Germany
Total R&D personnel in industry	1995	23,700	260,000
	1997	23,000	267,000
Breakdown: Percent of companies with ..... employees	1995/1997 [2]		
Less than 100		56.8	8.1
100 - 499		22.5	11.7
500 or more		20.7	80.3
R&D personnel intensity [1]			
Percentage total	1997	4.1	4.6
In companies with less than 100 employees	1995	6 - 7	1.8
Internal R&D expenditure in percent of turnover	1995	1.7	2.5
Number of patent applications [3]	1995	2,120	36,260
	1997	2,620	40,720

[1] Percentage of R&D personnel out of all personnel in mining and manufacturing.

[2] Former West Germany: 1995 based on SV data; new *Länder:* 1997 based on study by C. Hermann, et al. (1998).

[3] As defined here, Germany's new *Länder* include only the five territorial states. Figures for Berlin are for Berlin as a whole; these figures are allocated to former West Germany because West Berlin accounts for the majority of patent applications from Berlin.

Sources: SV-Wissenschaftsstatistik: Forschung und Entwicklung in der Wirtschaft 1995-1997; FuE Info 2/1998; German Patent Office, 1997 Annual Report; Federal Statistical Office, Fachserie 4, Reihe 4.1.1 and 4.1.2; Herrmann, C., Konack, T., Ständert, P. (1998), Analyse zur Entwicklung der Potentiale in Forschung und Entwicklung im Wirtschaftssektor in den neuen Bundesländern im Zeitraum 1990 bis 1997; NIW calculations and estimates.

5.3   Selected fields of technology – competitive positions at risk?

Although Germany's goods and technology portfolios generally match pretty well, there are still a number of striking differences. A consideration of these differences must take the following conditions into account:

• When foreign trade specialization is positive and technological specialization is negative, it can be assumed that problems could arise in the particular area.

• When technology specialization and foreign trade specialization are both negative, the questions arise: What is the long-term importance of this area? What is the economic importance of this field? Would it not be possible to pursue important options for the future?

- Furthermore, structural disruptions – such as sharp drops in production or job cuts – could begin to emerge in areas in which Germany has both technological and foreign trade specialization.

The following section examines whether "potential threats" to Germany's standing as an industrial location are beginning to develop in areas chosen according to these criteria, and what type of risks these are. When observing the case studies used here, it is interesting to consider the close connection between various technology fields – with the exception of office machines/EDP – and government demands on innovation activity (medical technology, pharmaceutical products, locomotives and rolling stock).

Office machines and computers

The "office machine and computer" field is one of Germany's most R&D-intensive industries: In Germany, nearly 15 percent of these products' production value is spent on R&D. These products account worldwide for some 20 percent of all patent applications submitted to the European Patent Office (EPO) in cutting-edge technology. They also constitute nine percent of the overall R&D-intensive sector. This industry is however extremely heterogeneous: The areas it covers range from simple office machines all the way up to ADP equipment (ranging from basic PCs to main frame computers) including parts and accessories.

The dominant role that the USA and Japan play at technological level is reflected by their relative patent shares, because no other country – with the exception of the Netherlands – reports above-average patent activity in this field. The USA and Japan generate not quite 75 percent of the patent applications made to the EPO (Table 5.5). These two countries however exchanged rankings between 1989/1990 and 1995/1996: While Japan's patent share shrank from 41 percent to 31 percent during this period, the USA's share expanded from 38 percent to 42.5 percent. Germany was able to gain some ground, increasing its share from seven percent to eight percent. In addition, a group of smaller countries, primarily from Southeast Asia, were able to increase their share from almost zero to three percent.

Taking another angle, the industry's heterogeneity is also evidenced by the fact that Japan and the USA (being the leading technology suppliers) account for large shares of the patent applications in this field whereas at the same time Southeast Asia's newly industrialized countries hold large shares of its global market, measured in terms of sales. Therefore, the one side (patent development) seems to have almost nothing to do with the other (world trade). This situation makes it very difficult to tally foreign trade structures with patent data. It is certain however that Germany currently does not have a quantitatively significant presence in this high-tech segment, particularly in the field of non-R&D-intensive technology.

Looking at future prospects for this sector, the following assessments are generally assumed:

- US and Japanese manufacturers are likely to continue dominating the new market for digital copiers for the time being.

- Non-European countries (Taiwan, Israel and even Japan) are exhibiting considerable momentum in the field of automatic data processing equipment as already clearly indicated by their increased patent application levels.

- Germany would have little to gain by trying to catch up in the area of basic hardware or standard PCs. However, the spread of computer technology and its links to other areas (such as to television and the automobile) open up a broad field in which Germany could compete: Namely, the skillful application of computer technology in traditional industries – in other words, integrating chips from the USA or Asia into products that are manufactured in Germany. This would appear to be a realistic prospect that would be reasonable given Germany's pattern of specialization.

Pharmaceutical products

Germany has been "the world's largest pharmacy" for years, with a world trade share of 20 percent. While it offers a wide range of medicines on the world market, it does not by contrast have a very strong presence on markets for active substances. Pharmaceutical innovation draws heavily on basic research findings, and the pharmaceutical industry is correspondingly integrated into collaborative structures at colleges and universities. However, when measured by the number of publications appearing in journals, pharmacology (with its 6.5-percent share of all German publications) is not one of Germany's strengths in the scientific realm.

When compared to the R&D expenditure reported by important competitors (USA, Japan, France and Great Britain), Germany's R&D expenditure on pharmaceutical products between 1980 and 1994 saw less-than-average growth, and even declined at times. In addition, these countries have also seen a visible shift in their R&D resources toward pharmaceutical products, whereas a similar focus has failed to develop in Germany. Accordingly, Germany has lost ground in the innovation field over the long term (active substances and application patents), while the USA in particular (with its 45-percent patent share) and, to an ever greater degree, Great Britain display strengths in this regard. Following a period of stagnation, the number of German patent applications in the pharmaceutical field has been on the rise again since 1993, in keeping with the international trend. Despite this, it appears that a potential threat to Germany's position in the pharmaceutical products field has built up.

Table 5.5:   Technology and foreign trade specialization in selected fields of technology

Field	GER	USA	JPN	FRA	GBR
Office machines/Computers					
Patent share in % - 1996	_9_	_41_	_30_	_4_	_5_
World trade share in %-1996	7	23	20	6	11
RCA 1996	-78	-22	14	-39	5
Thereof: Office machines (SITC 751)					
World trade share in % -1996	11	9	31	6	10
RCA 1996	-34	-119	111	-22	39
Thereof: EDP (SITC 752)					
World trade share in % -1996	7	22	16	7	12
RCA 1996	-82	-31	-20	-28	11
Thereof: Adv'ced parts f. computers (SITC 759)					
World trade share in % -1996	6	27	24	4	9
RCA 1996	-80	3	45	-64	-10
Pharmaceuticals					
Patent share in % - 1996	_12_	_45_	_12_	_5_	_8_
World trade share in % -1996	15	10	3	10	12
RCA 1996	16	31	-131	16	64
Thereof: Pharm. active ingredients (SITC 541)					
World trade share in % -1996	17	18	5	8	6
RCA 1996	6	57	-113	-26	6
Thereof: Medicaments (SITC 542)					
World trade share in % -1996	14	6	2	12	15
RCA 1996	23	0	-156	36	83
Medical diagnostic equipment					
Patent share in % - 1996	_13_	_44_	_10_	_3_	_8_
World trade share in % -1996	23	30	15	7	4
RCA 1996	75	81	17	1	6
Medical instruments					
Patent share in % - 1996	_13_	_50_	_5_	_5_	_5_
World trade share in % -1996	14	28	8	5	7
RCA 1996	5	89	-84	-59	27
Locomotives and rolling stock					
Patent share in % - 1996	_48_	_9_	_14_	_8_	_3_
World trade share in % -1996	15	13	3	8	3
RCA 1996	54	-6	0	125	50
Environmental technology					
Patent share in % - 1996	_25_	_22_	_16_	_9_	_7_
World trade share in % -1996	18	18	13	7	7
RCA 1996	49	56	55	8	32
Thereof:					
Waste World trade share in % -1996	17	14	17	6	7
RCA 1996	91	42	104	35	72
Water World trade share in % -1996	18	13	10	7	6
RCA 1996	24	30	58	-3	14
Air World trade share in % -1996	18	18	12	8	7
RCA 1996	58	53	69	27	32
M&C World trade share in % -1996	18	23	16	6	8
RCA 1996	43	72	31	-9	28

Sources: OECD: Foreign Trade by Commodities, CD-ROM; NIW calculations; EPAT, PCTPAT; FhG-ISI calculations.

The reason for this: The research and development of new pharmaceutical products has undergone dramatic technological change in recent years. This change has been driven by developments in the biotechnology and synthetic chemistry fields which are opening up doors to entirely new medicines for entirely new applications. As a result, the R&D process must draw not only on traditional science and technology fields such as chemistry, pharmacology and medical research in the future, but must also make increased use of know-how from the molecular biology, biochemistry, biotechnology, bioinformatics, combinatorial chemistry, automation and robotics fields. Germany exhibits only below-average technological specialization in those areas of biotechnology that are pertinent to pharmaceutical products. The portion of patent applications submitted to the EPO from throughout the world that were submitted for both the biotechnology field and the pharmaceutical products field has risen from 15 percent in 1989 to 20 percent in 1996. Looking at the USA, we even see a rise from 19 percent to 26 percent. Great Britain also built up pronounced strengths in this area whereas Germany reported only moderate growth during this period, expanding its share from 11 percent to 13 percent.[99]

German firms have responded to this, markedly increasing their R&D expenditure further since 1994. R&D spending has reached 10.5 percent of sales, while just under 12 percent of the total work force is involved in R&D activities. At the same time, the conditions for innovation have improved somewhat: The time frames for obtaining licenses for medicines have been shortened; reciprocal recognition of national licenses has been introduced within the EU; the time frames for receiving approval for bioengineering facilities have been shortened; pharmaceutical companies have stepped up their collaboration with German biotech firms; the number of biotech companies in Germany has doubled.[100] The pharmaceutical industry is even talking about a comeback for biotechnology in Germany, a field that was previously a popular focus of research conducted abroad. Germany's somewhat unfavorable image as a location for innovation has improved somewhat in the biotechnology field. Looking at technology levels however, Germany continues to trail the USA by a substantial distance.

Medical technology: Diagnostic equipment and instruments

Not quite eight percent of the sales volume generated in the medical technology and orthopedic technology industries is spent on R&D. Just under eight percent of the workers in this field are involved in conducting R&D.

---

[99]   The figures for 1996 are extrapolated.
[100] Cf.   Boston   Consulting   Group   (1998),   Innovationskraft: Forschende Arzneimittelhersteller am Standort Deutschland.

Medical **diagnostic equipment** is a highly interdisciplinary technological field that is developing along the interface between medical research and application, engineering sciences and various scientific disciplines. Although Germany has been quite successful on global markets in this field to date, it has not produced a particularly large number of patented inventions; in fact, a slightly negative trend can be observed. In light of the fact that the USA, presently the leading export nation in this field and responsible for approximately half of all patent applications being submitted, is further expanding its technological expertise as well and that Great Britain is developing into another technological rival, a potential threat to Germany's technological performance could develop in this area.

New findings being generated in the areas of material research, microsystem technology, information and communications engineering, optical technologies and biotechnology are likely to have a fundamental impact on the future development of medical diagnostic equipment. Consequently, expertise in these fields will be an important factor in Germany's future technological performance. All in all, Germany's positioning of its science and technology portfolio in these fields – with the exception of materials sciences – is generally unfavorable. However, translating new basic technological developments in individual disciplines into applications – actually one of German industry's strengths – will continue to be of critical importance in the future. The example of the microsystem technology field (Section 2.3.2) shows however that Germany still has a hard time in this area in particular.

As is the case of medical diagnostic equipment, there is also a marked discrepancy between Germany's technology and foreign trade specialization in **medical instruments**. Germany continues to report above-average export surpluses in this area despite its small patent shares. In the meanwhile this is also reflected in Germany's foreign trade statistics, with the country's share of global exports having fallen by one third to 13.5 percent in the years since 1986.

The USA clearly dominates both the patent statistics and the world markets. It accounted for some 43.5 percent of all EPO patent applications in this area in 1989/1990. This figure had grown to 49.5 percent by 1995/1996. During this period, all other countries saw their patent shares dwindle: Germany's share fell from 17 percent to 13.5 percent; Japan's share diminished from eight percent to six percent. However, both Switzerland and Sweden have good showings in their patent levels and their foreign trade balances.

Locomotives and rolling stock

The number of patents submitted in the area of railway and tramway locomotives and rolling stock has risen sharply worldwide – albeit along a low baseline – from not quite 200 applications in 1989 to some 350 in 1996. This increase is due first and foremost to growth in the number of patent applications from Germany. One contributing factor here was the integration of know-how from the former GDR. Germany – where R&D expenditure equaled six percent of turnover in 1995 and R&D personnel accounts for 12.7 percent of the total work force, both extremely high values – generated nearly 50 percent of all patent applications submitted to the EPO in this field. Japan, the USA, France and Switzerland follow Germany in ranking at a considerable distance. Germany's foreign trade balance in this field was also very positive for quite some time. Exports however fell by more than half in 1995/1996. Germany's relatively high export figures are not carved in stone. The German locomotive and rolling stock industry has increasing problems with sales. Canada (freight cars!) leads Germany, the USA, France, Switzerland and Italy as the world's largest exporter of locomotives and rolling stock. However, a large number of smaller countries such as South Korea and the Czech Republic are also successful players on the global market.

The pronounced discrepancies between patents and their realization on foreign markets can be attributed in part to the heterogeneity of this industry: While some segments are faced with tough competition over technology, others have to deal with fierce price competition. The large number of mergers evidences the fact that cost considerations are becoming increasingly important. Another reason for these discrepancies is that the market for long-distance locomotives and rolling stock is largely national. Further, governments frequently subsidize and promote prestige projects such as high-speed trains. Although regional preferences do not play as large a role in regard to trains used for local and urban transport purposes, the demand for rolling stock and the attendant technology is strongly influenced by factors governing their use such as design, maintenance and improvements which give local providers clear advantages on the market. Development and production are difficult to separate from one another.

The liberalization of rail transport in Germany – as a result of putting local transport into the hands of regional operators, or investing in local transport in eastern Germany and other European countries, for example – has led to growing markets, as well as to increasingly tough (international) competition.[101] Major suppliers[102] have their headquarters in Germany and their large market and output shares strongly influence Germany's ratios.

---

[101]   A decline in the number of employees indicates that many companies are facing this competition and are cutting their costs in preparation of price wars.
[102]   Adtranz and the Siemens Transportation Systems Group.

Appendix

Table A.1:　　　Expenditure on R&D and education in Germany 1992 – 1997

	In percent of GDP					
	1992	1993	1994	1995	1996 [1]	1997 [1]
(1) Research and development	2.5	2.4	2.3	2.3	2.3	2.4
Broken down by funding sector:						
Government	0.9	0.9	0.9	0.8	0.9	0.8
Private non-profit org.	0.0	0.0	0.0	0.0	0.0	0.0
Foreign	0.0	0.0	0.0	0.0	0.0	0.0
Corporate	1.5	1.5	1.4	1.4	1.4	1.5
Broken down by funding sector:						
Govern't; private non-profit org.	0.3	0.4	0.3	0.4	0.4	0.3
Universities	0.4	0.4	0.4	0.4	0.3	0.4
Firms	1.7	1.6	1.5	1.5	1.5	1.6
(2) Education and training [2]	5.3	5.5	5.3	5.5	5.5	5.4
(a) Government:	3.9	4.1	4.0	4.0	4.0	3.9
Schools, preschools [3]	2.8	2.9	2.8	2.9	2.9	2.8
Universities [3, 4]	0.5	0.5	0.5	0.5	0.5	0.5
Funding for the educa. system [3, 5]	0.2	0.2	0.2	0.2	0.2	0.2
Allowances for civil serv. [6]	0.4	0.4	0.4	0.4	0.4	0.4
(b) Firms [7]	1.0	1.0	0.9	1.0	1.0	1.0
(c) Private households [8]	0.4	0.5	0.5	0.5	0.5	0.5
(3) Cont. educa. and training [2, 14]	1.1	1.1	0.9	0.9	0.9	0.9
(a) Government:	0.7	0.6	0.5	0.5	0.5	0.5
Other educa. systems [3, 9]	0.1	0.1	0.1	0.1	0.1	0.1
Bundesanstalt für Arbeit [10]	0.4	0.4	0.3	0.3	0.3	0.3
Civil service [11]	0.1	0.1	0.1	0.1	0.1	0.1
(b) Firms [12]	0.3	0.3	0.3	0.3	0.3	0.3
(c) Private households [13]	0.1	0.1	0.1	0.1	0.1	0.1
Total from (1) + (2)	7.8	7.9	7.6	7.8	7.8	7.7
(1) + (2) + (3)	8.9	9.0	8.6	8.7	8.7	8.6

1. Education and training estimated on the basis of the education budget.
2. By sector providing the funding.
3. Basic funds excluding subsidies for civil servants.
4. Excluding R&D conducted by universities.
5. Financial assistance for students.
6. Plus an estimated DM 1.7 billion for allowances paid during times of illness.
7. Industry's net costs for initial training in the dual education system (BMBF projection) plus payments made by industry
   to universities, excluding R&D funds.
8. Revenues of public schools and preschools, and funding for the education system plus payments made to private schools,
   preschools and universities (estimated to be as high as government payments).
9. Adult education centers, libraries, academies.
10. Estimated to be 50% of the Bundesanstalt's (Federal Employment Service's) total spending on vocational education.
11. Estimated using the average value/employee in the companies (cf. footnote 12).
12. DM 10.2 billion estimated for further training courses in 1992/1993 & 1995 based on a corporate survey conducted by BIBB and
    Using a broader definition and including sick pay in its calculation, IW arrived at some DM 34 billion for 1995.
13. Estimated to be DM 3.6 billion for 1992 on the basis of GSOEP plus the revenues of public facilities in other education systems.
14. Expenditure solely for education. In other words, excluding sick pay and excluding subsistence payments by the Federal
    Service when individuals undergo further training. When opportunity costs are included, the results of this calculation are much higher.

Sources: DIW compilations, calculations and estimates based on data provided by SV-Wissenschaftsstatistik, the BMBF, Federal Statistical Office, the Bundesinstitut für Berufsbildung (BIBB = Federal Institute for Vocational Training) and the Institut der deutschen Wirtschaft (IW = Institute of the German Economy).

Table A.2:   Share of value added and employment generated by R&D-intensive
sectors in selected OECD countries 1973 – 1995

Sector	GER (former W. Germany)	USA	JPN	FRA	ITA	GBR
- Percentage of gross value added -						
R&D-intensive sectors						
1973-1975	13.3	9.7	12.3	9.2	9.2	11.0
1983-1985	15.0	9.5	13.3	8.8	8.1	9.5
1993-1995	12.2	8.5	11.5	7.7	6.4	8.0
Cutting-edge technology						
1973-1975	3.5	3.0	3.2	2.4	2.1	3.0
1983-1985	4.1	3.9	4.6	2.9	2.3	3.0
1993-1995	3.5	3.6	3.9	2.6	1.9	2.9
Advanced technology						
1973-1975	9.8	6.8	9.1	6.8	7.1	8.0
1983-1985	10.9	5.7	8.7	5.8	5.8	6.4
1993-1995	8.7	4.9	7.7	5.1	4.5	5.2
Non-R&D-intensive sectors						
1973-1975	22.3	13.8	20.7	17.2	18.6	16.2
1983-1985	16.3	10.8	16.1	13.4	16.3	11.9
1993-1995	13.7	9.5	13.5	11.7	13.9	10.2
Manufacturing						
1973-1975	35.5	23.5	33.0	26.4	27.7	27.2
1983-1985	31.3	20.4	29.4	22.1	24.4	21.3
1993-1995	25.9	18.0	25.0	19.3	20.4	18.3
- Percentage of all employees -						
R&D-intensive sectors						
1973-1975	14.3	8.5	9.3	10.2	8.2	12.6
1983-1985	14.1	7.6	9.5	9.0	6.6	9.5
1993-1995*	12.8	5.9	9.3	7.5	5.8	8.2
Cutting-edge technology						
1973-1975	3.7	3.1	2.7	2.6	1.6	3.3
1983-1985	3.5	3.5	3.4	2.6	1.4	2.9
1993-1995*	3.1	2.5	3.1	2.3	1.4	2.7
Advanced technology						
1973-1975	10.6	5.4	6.6	7.5	6.5	9.3
1983-1985	10.6	4.1	6.0	6.5	5.1	6.6
1993-1995*	9.7	3.4	6.2	5.2	4.3	5.4
Non-R&D-intensive sectors						
1973-1975	21.7	14.5	17.2	16.6	19.2	18.7
1983-1985	17.8	11.2	14.6	13.6	16.6	13.5
1993-1995*	14.9	9.7	13.8	11.0	14.7	11.4
Manufacturing						
1973-1975	36.0	23.0	26.4	26.8	27.4	31.3
1983-1985	31.8	18.8	24.1	22.6	23.1	23.1
1993-1995*	27.7	15.5	23.2	18.5	20.4	19.6

* USA: 1993-1994; GBR: 1993.

Cutting-edge technology: Pharmaceutical products; computers/office machines; radio/TV/telecommunications; aircraft and spacecraft; precision instruments;
    optical products/watches & clocks.
Advanced technology: Other chemical products, mechanical engineering products; electrical engineering products w/o radio/TV/telecommunications; locomotives
    and rolling stock; motor vehicles.

Sources: OECD: STAN-Database; Economic Outlook; DIW calculations and estimates.

Table A.3:   Germany's specialization in R&D-intensive goods 1991 – 1996[1] (RCA values) and Germany's foreign trade in R&D-intensive goods – 1996

Product group	SITC	\multicolumn{6}{c}{RCA value}						Exports in DM bil. 1996	Imports in DM bil. 1996
		1991	1992	1993	1994	1995	1996		
Total R&D-intensive goods		21	24	23	24	25	24	369.3	230.1
Cutting-edge technology		-11	-10	-14	-12	-8	-11	108.8	95.9
Radioactive	(525)	-25	-124	-136	-162	-146	-120	0.3	0.8
Nucl., water, windpower	(718)	93	83	74	107	90	55	1.9	0.8
Of: Chemical industry		36	36	40	43	34	32	20.2	11.6
Advanced organic chemicals	(516)	12	6	4	34	39	42	2.6	1.3
Pharmacological active ingredients	(541)	34	35	37	33	20	8	6.1	4.4
Advanced plastics	(575)	39	39	41	37	30	36	8.8	4.9
Herbicides, etc.	(591)	61	67	101	117	87	79	2.7	1.0
EDP	(752)	-77	-95	-89	-86	-81	-86	10.6	19.8
Telekommunications equipment	(764)	-21	-18	-21	-14	-4	7	14.3	10.6
Of: Electrical engineering		5	9	-2	-11	-11	-13	30.2	27.2
Medical electronics	(774)	76	73	83	87	78	72	4.1	1.6
Semi-conductor devices	(776)	-26	-38	-42	-42	-42	-47	11.1	14.0
Advanced electrical engineering	(778)	14	27	11	-1	2	2	15.1	11.6
Aircraft and spacecraft		-34	-28	-37	-33	-16	-31	16.2	17.4
Turbines and reaction engines	(714)	5	8	-28	-34	-59	-46	4.2	5.2
Aircraft	(792)	-40	-34	-38	-33	-4	-26	12.0	12.2
Of: Measuring and control technology		37	39	48	44	49	45	14.7	7.4
Optical instruments	(871)	29	25	40	37	43	36	1.2	0.6
Measuring and control technology	(874)	38	41	48	45	49	46	13.6	6.8
Arms and ammunition	(891)	*	*	*	*	*	*	*	*
Advanced technology		37	42	44	46	43	43	260.5	134.1
Of: Chemical industry		46	43	48	42	38	39	47.0	25.2
Synthetic fibers	(266)	66	66	67	59	47	55	1.1	0.5
Heterocyclic chemistry	(515)	36	35	45	26	20	21	5.6	3.6
Rare anorganic materials	(522)	46	34	31	30	29	28	3.0	1.8
Other anorganic materials	(524)	7	-6	3	15	-9	-10	0.7	0.6
Synthetic dyes	(531)	140	135	126	125	108	98	3.7	1.1
Pigments, paints, varnishes	(533)	78	72	77	81	78	85	5.8	1.9
Medicaments	(542)	30	30	38	26	22	23	10.1	6.3
Essential oils, perfume, flavor	(551)	2	7	12	8	10	12	0.9	0.7
Polyethers and resins	(574)	22	20	22	37	41	39	3.8	2.1
Other advanced chemicals	(598)	60	56	65	57	56	55	9.0	4.1
Photographical chemicals	(882)	-5	-2	-2	2	-3	5	3.4	2.5

- Continued on next page -

Product group	SITC	RCA value						Exports in DM bil. 1996	Imports in DM bil. 1996
		1991	1992	1993	1994	1995	1996		
Of: Machines for particular industr.		113	119	131	129	140	140	41.1	8.0
Textile and leather machines	(724)	153	163	181	170	171	167	9.0	1.3
Paper & pulp machinery	(725)	82	82	97	87	109	106	3.0	0.8
Printing & pulp machinery	(726)	126	126	140	137	149	161	7.2	1.1
Food processing machines	(727)	93	84	103	124	138	145	2.1	0.4
Advanced machine tools	(728)	102	108	111	115	130	129	19.8	4.3
Metal processing machines		58	65	90	89	80	76	11.6	4.3
Metal machine tools	(731)	52	66	96	93	75	68	4.9	2.0
Cermet-machine tools	(733)	89	99	123	122	115	126	2.2	0.5
Machine tool components	(735)	24	21	47	51	50	44	1.9	1.0
Advanced machine tools	(737)	77	73	89	90	93	89	2.5	0.8
Of: Other machinery & equipm.		62	56	63	65	65	65	26.2	10.8
Heating and cooling equipment	(741)	58	43	43	49	45	41	7.5	3.9
Mechanical handling equipment	(744)	49	44	49	54	65	73	6.9	2.6
Other non-electrical machines	(745)	100	101	115	115	114	115	8.7	2.2
Ball and roller bearings	(746)	24	24	31	22	21	16	3.1	2.1
Office machines		-53	-56	-72	-73	-67	-71	7.9	12.7
Office machines, word processing	(751)	-39	-32	-43	-43	-35	-34	1.8	2.0
Advanced parts for computers	(759)	-58	-63	-81	-81	-76	-79	6.1	10.7
Of: Telecommunications		-85	-83	-106	-95	-90	-93	3.9	7.7
TV and video equipment	(761)	-56	-78	-98	-77	-76	-93	1.4	2.7
Radio broadcast	(762)	-132	-132	-157	-157	-147	-125	0.9	2.4
Sound recording and reproducing	(763)	-134	-57	-81	-74	-63	-70	1.7	2.6
Of: Electrical engineering		42	38	37	31	33	32	21.4	12.3
Traditional electronics	(772)	63	66	63	59	58	57	16.6	7.4
Optical fibers and other cable	(773)	-17	-30	-29	-37	-27	-26	4.8	4.9
Of: Automobile construction		27	42	48	59	50	49	87.9	42.5
Passenger cars	(781)	29	45	50	62	52	52	78.1	36.7
Commercial vehicles	(782)	15	19	30	34	36	29	9.9	5.8
Railroad vehicles	(791)	134	152	125	127	84	54	1.4	0.7
Of: Measuring & control technology		34	29	26	19	15	11	5.1	3.6
Medical instruments	(872)	30	23	21	14	9	6	4.1	3.1
Traditional measuring instruments	(873)	58	60	56	42	41	34	1.0	0.6
Of: Photographical & optical products		-15	-17	-17	-20	-16	-24	3.3	3.3
Photograph. apparatus & equipment	(881)	-18	-20	-19	-27	-25	-41	1.4	1.7
Optical fibers, contact lenses	(884)	-12	-14	-16	-13	-7	-9	1.9	1.6
Other advanced technol.		15	-1	5	0	-9	-5	3.6	3.0
Advanced industrial abrasives	(277)	-65	-72	-25	-15	-39	-16	0.2	0.1
Precious NF base metals	(689)	-73	-82	-66	-67	-72	-79	0.3	0.5
Advanced fine ceramics	(663)	31	13	15	12	4	7	3.1	2.3

[1.] 1995 and 1996 = data from the Federal Statistical Office.   * Inapplicable data.

A positive value indicates that the respective product group's export-import ratio is higher than the export-import ratio for total manufactured goods.

Sources: OECD and Federal Statistical Office: Foreign Trade By Commodities; unpublished analyses; NIW calculations.

Table A.4:   Export quotas, percentage change in foreign and domestic sales for all R&D-intensive industries in Germany 1995 - 1997 and 1st half of 1998

WZ93 Sector	Export quota in % 1997	Average annual change 1995 - 1997		Change 1st halves 97/98
- Germany -		Foreign sales in %	Domestic sales in %	Foreign sales in %
CET Cutting-edge technology*	**47.9**	**17.4**	**2.6**	**11.7**
23.30 Nuclear fuel	.	.	.	.
24.20 Pesticides & other agro-chemical products	73.3	12.1	-7.6	9.5
24.41 Basic pharmaceutical products	78.6	12.8	9.4	20.8
24.42 Pharmaceut. preparations	36.3	10.0	-1.2	8.5
29.60 Weapons and ammunition	35.1	36.0	2.9	-4.8
30.02 Computers & other information processing equipment	34.5	9.6	7.7	13.8
32.10 Electronic valves and tubes & other electronic components	62.2	14.0	6.7	12.7
32.20 TV & radio transmitters & apparatus for line telephony	51.4	37.8	2.1	8.6
33.20 Instr. & appliances for measur'g, check'g, test'g, navig.	40.4	7.1	-1.0	18.6
35.30 Aircraft and spacecraft	63.2	25.4	13.0	11.8
AT Advanced technology	**48.2**	**9.6**	**0.7**	**11.7**
24.11 Industrial gases	2.9	-0.2	-0.2	-8.9
24.12 Dyes and pigments	72.6	7.7	-5.7	-2.3
24.13 Other inorgan. basic chemicals	50.4	5.3	-0.9	1.9
24.14 Other organic basic chemicals	53.8	3.7	2.3	0.6
24.16 Plastics in primary forms	55.5	8.2	-0.4	3.2
24.30 Paints, varnishes & similar coatings, printing ink and mastics	28.2	10.5	0.5	10.0
24.52 Perfumes & toilet preparations	24.1	-1.1	-3.8	12.0
24.63 Essential oils	55.8	9.2	-5.1	-8.6
24.64 Photograph. chemical material	73.7	-0.3	-10.1	7.0
24.66 Other chemical products, n.e.c.	56.6	10.4	-1.4	6.3
24.70 Man-made fibers	69.6	6.3	-6.7	-4.7
26.23 Ceramic insulators & insulating fittings	48.7	5.1	-7.1	-0.7
28.30 Steam generators,except central heating hot water boilers	15.4	0.1	4.5	-28.8
28.62 Tools	32.4	7.5	2.3	14.0
29.11 Engines & turbines, except aircraft, vehicle & cycle engines	66.0	14.6	-0.7	47.0
29.12 Pumps and compressors	47.5	3.5	-4.4	10.4
29.13 Taps and valves	31.3	6.0	-0.6	9.7
29.14 Bearings, gears, gearing & driving elements+B73	41.5	6.6	2.0	6.5
29.21 Furnaces and furnace burners	42.2	2.9	-0.2	-4.4
29.22 Lifting & handling equipment	35.1	11.3	-3.4	18.8
29.23 Non-domestic cooling and ventilation equipment	31.4	6.2	1.0	6.7
29.24 Other general purpose machinery n.e.c.	46.4	8.0	1.8	10.5
29.40 Machine-tools	48.6	3.2	0.1	14.5
29.51 Machinery for metallurgy	64.1	-4.5	11.7	20.6
29.52 Machinery for mining, quarrying and construction	55.1	11.6	-11.3	33.1
29.53 Machinery for food, beverages and tobacco processing	59.6	6.2	-10.0	-5.4
29.54 Machinery for textile, apparel and leather production	78.8	3.0	-6.5	21.0
29.55 Machinery for paper and paperboard production	64.2	12.3	0.2	3.5
29.56 Other special purpose machinery n.e.c.	50.7	15.7	7.7	12.1
30.01 Office machinery	52.5	1.4	-8.0	14.1
31.10 Electric motors, generators and transformers	35.8	0.7	-4.6	24.2
31.20 Electricity distribution and control apparatus	34.1	10.3	0.4	11.4
31.30 Insulated wire and cable	32.3	37.3	5.1	0.6
31.40 Accumulators, primary cells and primary batteries	37.1	2.7	4.6	3.9
31.50 Lighting equipment and electric lamps	35.3	9.1	-3.1	4.6
31.61 Electrical equipment for engines and vehicles n.e.c.	29.9	4.7	8.0	6.8
31.62 Other electrical equipment n.e.c.	33.2	1.4	-1.7	14.0
32.30 TV & radio receivers, sound/video record. or reproduc. app.	35.0	-23.0	-6.2	14.9
33.10 Medical & surgical equipment and orthopaedic appliances	44.8	10.7	1.7	13.1
33.40 Optical instruments and photographic equipment	49.4	6.8	-4.6	9.0
34.10 Motor vehicles	58.0	16.0	3.4	14.7
35.20 Railway & tramway locomotives and rolling stock	24.0	6.1	10.2	-16.7
RD R&D-intensive industries	**48.1**	**10.9**	**1.1**	**11.7**
NRI Non-R&D-intensive industries	**21.9**	**5.1**	**1.4**	**9.5**
MI Total manufacturing industry	**34.0**	**8.8**	**1.3**	**10.9**

* Without nuclear fuel.

Sources: Federal Statistical Office: statistics on the manufacturing industry; NIW calculations.

Table A.5:   Change in net output during the upswing (1993 - 1997) by branch of
industry in Germany

- Germany, kind-of-activity-units, WZ 93 - new list		
Average annual change in % 1993 - 1997	**Cutting-edge technology**	**Advanced technology**
>3.1 (> Average for all R&D-intensive industries)	Electronic valves & tubes & other electronic components Computers & other information processing equipment TV & radio transmitters and apparatus for line telephony Pharmaceutical preparations Pesticides and other agro-chemical products	Plastics in primary forms Machinery for metallurgy Steam generators, except central heating hot water boilers Other special purpose machinery n.e.c. Bearings, gears, gearing and driving elements Essential oils Motor vehicles Dyes and pigments Other organic basic chemicals Tools Man-made fibers Electric motors, generators and transformers Other chemical products n.e.c. Industriral gases Ceramic insulators and insulating fittings Engines & turbines, except aircraft, vehicle and cycle engines Optical instruments and photographic equipment
2.2 - 3.1 (Still > overall industry average)		Pumps and compressors Machine-tools Electrical equipment for engines and vehicles
0 - <2.2 (< Overall industry average, but still positive rate of change)	Instruments & appliances for measuring, checking, testing and navigation Weapons and ammunition	Machinery for paper and paperboard production Paints, varnishes and similar coatings, printing ink and mastics Electricity distribution and control apparatus Other inorganic basic chemicals Insulated wire and cable Other electrical equipment n.e.c. Lifting and handling equipment Other electrical equipment n.e.c. Machinery for textile, apparel and leather production Medical & surgical equipment and orthopaedic appliances Taps and valves
<0 (negative rate of change)	Pharmaceutical preparations Aircraft and spacecraft	Office machinery Other general purpose machinery n.e.c. Accumulators, primary cells and primary batteries Furnaces and furnace burners Photographic chemical material Lighting equipment and electric lamps Machinery for food, beverages and tobacco processing Machinery for mining, quarrying and construction Perfumes and toilet preparations Railway & tramway locomotives and rolling stock TV & radio receivers, sound/video recording or reproducing equipment

Sources: Federal Statistical Office, Fachserie 4, Reihe 2.1; NIW calculations.

Table A.6:   Change in employment levels in Germany according to branch of industry 1995 – 1997

- Germany, kind-of-activity-units, WZ 93 - new list		
Average annual change in % 1995 - 1997	Cutting-edge technology	Advanced technology
>0 (positive change)	Weapons and ammunition Basic pharmaceutical	Other special purpose machinery n.e.c. Medical & surgical equipment & orthopaedic appliances Motor vehicles
0 - >-3.2 (negative change, but > all R&D-intensive industries)	Pesticides and other agro-chemical products Instruments & appliances for measuring, checking, testing, navigating Television and radio transmitters and apparatus for line telephony	Other organic basic chemicals Tools Bearings, gears, gearing and driving elements Essential oils Paints, varnishes & similar coatings, printing ink & mastics Industrial gases Ceramic insulators and insulating fittings Engines and turbines (except aircraft, vehicle and cycle engines) Electrical equipment for engines and vehicles n.e.c. Other chemical products n.e.c. Machine-tools Taps and valves Insulated wire and cable Other general purpose machinery n.e.c. Machinery for paper and paperboard production
<-3.2 (< Average for all R&D-intensive industries)	Pharmaceutical preparations Electornic valves and tubes and other electronic components Aircraft and spacecraft Computers and other information processing Nuclear fuel	Furnaces and furnace burners Plastics in primary forms Perfumes and toilet preparations Electricity distribution and control apparatus Non-domestic cooling and ventilation equipment Lighting equipment and electric lamps Lifting and handling equipment Dyes and pigments Other inorganic basic chemicals Pumps and compressors Steam generators, except central heat'g hot water boilers Other electrical equipment n.e.c. Machinery for mining, quarrying and construction Office machinery Man-made fibers Machinery for food, beverages and tobacco processing Machinery for textile, apparel and leather production Electric motors, generators and transformers Optical instruments and photographic equipment Accumulators, primary cells and primary batteries Railway & tramway locomotives and rolling stock Machinery for metallurgy TV & radio receivers, sound/video recording or reproducing app. Photographic chemical material

Sources: Federal Statistical Office, Fachserie 4, Reihe 4.1.1; NIW calculations.

Table A.7:   Service intensity[1] in German industry 1990 – 1997

Sector	Former West Germany		Germany	
	1990	1996	1996	1997
Mechanical engineering	35.8	39.8	39.5	39.7
Motor vehicles, motors	23.6	26.7	26.6	26.9
Aircraft	59.1	56.5	56.2	55.8
Electrical engineering	39.1	44.9	44.0	45.2
Optical products	44.0	51.5	52.6	53.0
Precision instruments	30.9	32.8	32.3	32.6
Chemical products	48.6	53.4	53.1	53.7
Office machines	51.1	62.9	62.3	63.8
EDP equipment	71.0	75.0	73.9	77.3
Total manufacturing industry	32.4	36.0	35.2	35.5

[1] Percentage of employees out of total employees.

Source: Federal Statistical Office: statistics on employees who are liable to social
          security; NIW calculations.

Table A.8:   Human capital intensity[1] among services in Germany 1990 – 1997

Sector	Former West Germany		Germany	
	1990	1996	1996	1997
Mechanical engineering	16.6	20.0	22.7	23.0
Motor vehicles	26.3	32.0	32.6	33.1
Aircraft	37.0	36.2	36.6	37.1
Electrical engineering	26.2	29.8	31.1	31.9
Optical products	14.4	14.8	18.5	19.4
Precision tools	13.5	15.2	16.0	16.7
Chemicals	19.5	22.2	23.7	24.1
Office machines	8.5	7.4	11.0	11.6
EDP equipment	29.8	29.7	30.1	31.2
Total manufacturing industry	14.9	17.0	18.3	18.9

[1] Percentage of university/technical college graduates out of all employees.

Sources: Federal Statistical Office: statistics on employees who are liable to social
          security; NIW calculations.

Table A.9:   Percentage of qualified employees[1] in Germany   1990 – 1997

Sector	Former West Germany		Germany	
	1990	1996	1996	1997
Energy, water, mining	74.9	80.7	79.0	80.0
Manufacturing industry	64.2	69.9	71.5	71.7
Including				
Mechanical engineering	75.6	81.2	82.1	82.1
Motor vehicles	66.0	74.8	75.1	75.6
Aircraft	87.1	88.8	88.4	88.5
Electrical engineering	64.2	71.5	72.9	73.4
Optical products	68.1	74.2	75.8	76.0
Precision instruments	68.3	71.4	71.7	72.0
Chemicals	70.7	76.3	77.1	77.5
Office machines	60.2	69.7	71.5	72.6
EDP equipment	73.5	78.2	78.3	78.6
Construction industry	68.3	66.7	70.2	70.2
Trade	72.8	74.2	75.1	74.9
Transport/Communications	66.4	69.4	73.0	72.7
Banking, insurance	80.0	84.3	84.5	85.0
Services n.e.c.	67.5	69.4	71.1	70.6
Including				
Business services	72.7	72.6	73.6	72.7
Trade and industry *	67.6	71.1	72.6	72.5

[1] Percentage of employees who have completed formal vocational training out of all employees.

* Excluding agriculture and government.

Sources: Federal Statistical Office: statistics on employees who are liable to social security; NIW calculations.

Table A.10:   Percentage of highly qualified personnel[1] in Germany   1990 – 1997

Branch	Former West Germany		Germany	
	1990	1996	1996	1997
Energy, water, mining	6.9	8.9	10.1	10.4
Manufacturing industry	4.8	6.1	6.4	6.7
Including				
Mechanical engineering	5.9	8.0	9.0	9.2
Motor vehicles	6.2	8.5	8.7	8.9
Aircraft	21.9	20.4	20.6	20.7
Electrical engineering	10.2	13.4	13.7	14.4
Optical products	6.3	7.6	9.7	10.3
Precision instruments	4.2	5.0	5.2	5.4
Chemicals	9.5	11.9	12.6	12.9
Office machines	4.3	4.6	6.9	7.4
EDP equipment	21.1	22.3	22.2	24.1
Construction industry	2.1	2.5	2.9	2.9
Trade	2.4	3.2	3.4	3.5
Transport/Communications	1.5	1.9	2.8	2.8
Banking, insurance	5.9	8.1	8.7	9.1
Services n.e.c.	10.2	11.4	12.7	13.0
Including				
Business services	11.8	15.1	15.6	15.9
Trade and industry *	5.4	6.8	7.4	7.7

[1] Percentage of university/technical college graduates out of all employees.

* Excluding agriculture and government.

Sources: Federal Statistical Office: statistics on employees who are liable to social security; NIW calculations.

List A.1:   NIW ISI List of R&D-intensive goods according to SITC III

SITC III	Short description (abridged version of official terminology)
Cutting-edge technology:	
516	Advanced organic
525 *	Radioactive
541	Pharmacological active
575	Advanced plastics
591	Herbicides, etc.
714 *	Turbines and reaction
718 *	Nuclear, water, wind power
752	ADP
764	Telecommunications
774	Medical electronics
776	Semi-conductor devices
778	Advanced electrical
792 *	Aircraft and spacecraft
871	Optical instruments
874	Measuring and control
891 *	Arms and ammunition
Advanced technology:	
266	Synthetic fibers
277	Advanced industrial
515	Heterocyclic chemistry
522	Rare anorganic
524	Other anorganic
531	Synthetic dyes
533	Pigments, paints, varnishes
542	Medicaments
551	Essential oils, perfume,
574	Polyethers and resins
598	Other advanced
663	Advanced fine ceramics
689	Precious non-ferrous base
724	Textile and leather
725	Paper and pulp
726	Printing and bookbinding
727	Industrial food-processing
728	Advanced machine-
731	Metal machine tools
733	Cermet machine tools
735	Machine tool
737	Other machine tools
741	Heating and cooling
744	Mechanical handling
745	Other non-electrical
746	Ball and roller
751	Office machines, word processing
759	Advanced computer
761	Television and video
762	Radio-broadcast, radiotelephony
763	Sound recording and
772	Traditional electronics
773	Optical fibers and other
781	Passenger cars
782	Commercial vehicles
791	Railroad vehicles
872	Medical instruments
873	Traditional measuring
881	Photographic apparatus and
882	Photographical chemicals
884	Optical fibers, contact lenses

"Cutting-edge technology" covers goods with an R&D intensity of more than 8.5% of turnover. "Advanced technology" covers goods with an R&D intensity of more than 3.5%, but less than 8.5% of turnover. This distinction does not imply any "judgement": These two groups differ significantly in terms of their R&D intensity on the one hand, and by the degree of protection they are granted on the other. Cutting-edge technology goods are those with the highest R&D intensity and are frequently subject to intervention in the form of subsidies, procurement and/or import protection (marked by *).

Source: Grupp/Legler, Innovationspotential und Hochtechnologie. Report by FhG-ISI, NIW and Gewiplan for the BMFT (1991)

List A.2:   List of R&D-intensive branches of industry

1996 version; Revision of the 1990 NIW/ISI list, based on SITC / WZ93

WZ93	Activity

Cutting-edge technology

23.30	Processing of nuclear fuel
24.20	Manufacture of pesticides and other agro-chemical products
24.41	Manufacture of basic pharmaceutical products
24.42	Manufacture of pharmaceutical preparations
29.60	Manufacture of weapons and ammunition
30.02	Manufacture of computers and other information processing equipment
32.10	Manufacture of electronic valves and tubes and other electronic components
32.20	Manufacture of television and radio transmitters and apparatus for line telephony and line telegraphy.
33.20	Manufacture of instruments and appliances for measuring, checking, testing, navigating & other purposes
35.30	Manufacture of aircraft and spacecraft

Advanced technology

24.11	Manufacture of industrial gases
24.12	Manufacture of dyes and pigments
24.13	Manufacture of other inorganic basic chemicals
24.14	Manufacture of other organic basic chemicals
24.16	Manufacture of plastics in primary-forms
24.30	Manufacture of paints, varnishes and similar coatings, printing ink and mastics
24.52	Manufacture of perfumes and toilet preparations
24.63	Manufacture of essential oils
24.64	Manufacture of photographic chemical material
24.66	Manufacture of other chemical products n.e.c.
24.70	Manufacture of man-made fibers
26.23	Manufacture of ceramic insulators and insulating fittings
28.30	Manufacture of steam generators, except central heating hot water boilers
28.62	Manufacture of tools
29.11	Manufacture of engines and turbines, except aircraft, vehicle and cycle engines
29.12	Manufacture of pumps and compressors
29.13	Manufacture of taps and valves
29.14	Manufacture of bearings, gears, gearing and driving elements
29.21	Manufacture of furnaces and furnace burners
29.22	Manufacture of lifting and handling equipment
29.23	Manufacture of non-domestic cooling and ventilation equipment
29.24	Manufacture of other general purpose machinery n.e.c.
29.40	Manufacture of machine-tools
29.51	Manufacture of machinery for metallurgy
29.52	Manufacture of machinery for mining, quarrying and construction
29.53	Manufacture of machinery for food, beverage and tobacco processing
29.54	Manufacture of machinery for textile, apparel and leather production
29.55	Manufacture of machinery for paper and paperboard production
29.56	Manufacture of other special purpose machinery n.e.c.
30.01	Manufacture of office machinery
31.10	Manufacture of eletric motors, generators and transformers
31.20	Manufacture of electricity distribution and control apparatus
31.30	Maufacture of insulated wire and cable
31.40	Manufacture of accumulators, primary cells and primary batteries
31.50	Manufacture of lighting equipment and electric lamps
31.61	Manufacture of electrical equipment for engines and vehicles n.e.c.
31.62	Manufacture of other electrical equipment n.e.c.
32.30	Manufacture of tv and radio receivers, sound or video recording or reproducing apparatus & associated goods.
33.10	Manufacture of medical and surgical equipment and orthopaedic appliances [2]
33.40	Manufacture of optical instruments and photographic equipment
34.10	Manufacture of motor vehicles
35.20	Manufacture of railway and tramway locomotives and rolling stock [1]

[1], [2]: This list of R&D-intensive industries is generally based on the 1990 NIW/ISI list that is highly disaggregated according to product groups (Cf. Grupp/Legler, Innovationspotential und Hochtechnologie. Bericht des FhG-ISI, des NIW und Gewiplan an den BMFT (1991). This list has been reorganized to reflect the WZ 93 system. In addition, apparent shifts in industrial research priorities have been taken into account (in other words, several product groups have been added (1) because their R&D intensity increased markedly during the 1990s and a branch of industry (2) has been reassigned to the advanced technology field because of new findings. (Cf. Gehrke et al., Materialien des NIW zur Berichterstattung zur technologischen Leistungsfähigkeit Deutschlands 1996 (published Jan. 1997)).

List A.3:   NIW List of know-how-intensive sectors

Know-how-intensive sectors (Producing industry)		Know-how-intensive sectors (Services)	
WS	Sector, business group	WS	Sector, business group
04	Electricity, gas, steam and hot water supply	621	Wholesale trade and commission trade
051	Mining of lignite	625	Other retail trade
06	Mining of metal ores	630	Deutsche Bundesbahn (railway)
07	Crude petroleum and natural gas	64	Deutsche Bundespost (post office, telephone, telegraph)
09	Basic chemicals	66	Water transport, harbors
10	Man-made fibers	68	Air transport, airports, other transport
11	Refined petroleum products	69	Financial intermediation, insurance
231	Tanks, reservoirs and containers of metal	722	Chimney sweeps
24	Railway or tramway rolling stock	731	Beauty treatment, pedicure
260	Metal processing machines	74	Higher education, general and vocational schools
261	Smelting & rolling mill equip., construction machinery	750	Independent teachers
264	Machines for food, beverages and tobacco industry	751	Other educational institutions operated by non-profit organizations
266	Wood processing machines	752	Other educational institutions operated by territorial authorities
267	Paper and printing machinery	753	Private reform schools
268	Laundry machinery,footwear and leather machinery	754	Reform schools operated by non-profit organizations
27	Gearwheels, gears, bearings, etc.	755	Reform schools operated by territorial authorities
280	Motor vehicles and motors	760	Private theaters
31	Shipbuilding	761	Theaters operated by non-profit organizations
32	Aircraft	762	Theaters operated by territorial authorities
33	Office machinery, automatic data processing	763	Cinemas
340	Electrical engineering	764	Radio and television stations
341	Accumulators, primary cells and primary batteries	770	Publishing
342	Power current equipment	772	Libraries operated by non-profit organizations
343	Generators	773	Libraries operated by territorial authorities
344	Wire and cable	774	New agencies
345	Electrical appliances and housewares	78	Independent health-care services
346	Lighting equipment and electric lamps	79	Legal, accounting, book-keeping and auditing activities
347	Television and radio receivers	80	Architectural and engineering activities & related technical consultancy
348	Measuring and control technology, telecommunications	81	Real estate activities
35	Precision instruments, optics	82	Advertising, exhibitions
373	Arms and ammunition	862	Typing pools
580	Tobacco products		

Know-how-intensive production is indicated by an above-average share of university and other tertiary level graduates, highest-paid white-collar workers and skilled workers.

Source: Revision of the list of know-how-intensive sectors from Gehrke et al., Wissensintensive Wirtschaftszweige und ressourcenschonende Technik 1995, by NIW.

Abbreviations

AUT	Austria
BEL	Belgium
BIBB	Bundesinstitut für Berufsbildung (Federal Institute for Vocational Training)
bil.	Billion
BMBF	Federal Ministry of Education and Research, Bonn
BMFT	Federal Ministry of Research and Technology, Bonn
CAN	Canada
CEEC	Central and East European countries
DIW	Deutsches Institut für Wirtschaftsforschung (German Institute for Economic Research), Berlin
DM	Deutsche Mark
GPO	German Patent Office, Munich
EDP	Electronic data processing
EU	European Union
EPO	European Patent Office
EPAT	Database of the European Patent Office
ESP	Spain
FhG	Fraunhofer Gesellschaft, Munich
FhG-ISI	Fraunhofer Institute for Systems and Innovation Research, Karlsruhe
FIN	Finland
FRA	France
R&D	Research and development
GATT	General Agreement on Tariffs and Trade
GBR	Great Britain and Northern Ireland
GDP	Gross domestic product
GER	Germany

HGF	Hermann von Helmholtz-Gemeinschaft Deutscher Forschungszentren (Hermann von Helmholtz Association of National Research Centres)
HHI	Heinrich-Hertz-Institut für Nachrichtentechnik (Heinrich Hertz Institute), Berlin
HS	Harmonized system of foreign trade statistics
IAB	Institut für Arbeitmarkt und Berufsforschung (Institute for Employment Research)
ifo	ifo Institut für Wirtschaftsforschung e.V. (ifo Institute for Economic Research), Munich
ILO	International Labour Organization, Geneva
IMD	International Institute for Management Development, Lausanne
ITA	Italy
I&C	Information and communication
IW	Institut der deutschen Wirtschaft (Institute of the German Economy), Cologne
PN	Japan
Mil.	Million
MPS	Max Planck Society
n.a.	Not available
NAS	National accounting system
n.e.c.	Not elsewhere classified
NED	Netherlands
NIW	Niedersächsisches Institut für Wirtschaftsforschung (Institute for Economic Research of Lower Saxony), Hanover
NL	New *Länder*
OECD	Organisation for Economic Cooperation and Development, Paris
PATDPA	Patent database of the German Patent Office
PPP	Purchasing power parity

RCA	Revealed Comparative Advantage: A positive value indicates that the export-import ratio for this product group is higher than the export-import ratio for total manufactured goods.
R&D	Research and development
RPS	Relative Patent Share: A positive value indicates that the share of patents in this field is larger than for total patents.
RWA	Relative World Trade Share: A positive value indicates that the share of world trade for this product group is larger than for total manufactured goods.
SCI	Science Citation Index
SITC	Standard International Trade Classification
SUI	Switzerland
SV	Stifterverband für die Deutsche Wissenschaft (Donors' Association for the Promotion of Sciences and Humanities in Germany), Essen
SWE	Sweden
UNESCO	United Nations Educational Scientific and Cultural Organization, New York
USA	United States of America
WGL	Wissensgemeinschaft Gottfried Wilhelm Leibniz ("Blue List" institutes, Leibniz Institutes)
WSV	Gemeinnützige Gesellschaft für Wissenschaftsstatistik des Stifterverbandes für die Deutsche Wissenschaft (Society for Economic Statistics of the Donors' Association for the Promotion of Sciences and Humanities in Germany), Essen
WZ	Classification for industrial branches
ZEW	Zentrum für Europäische Wirtschaftsforschung (Center for European Economic Research), Mannheim
$	US dollar

List of Charts

List of Tables

Bibliography

Becher, G., Gering, Th., Lang, O. and U. Schmoch (1996): Themenkreis Patentwesen und Hochschulen, Studie für das BMBF. Karlsruhe und Basel.

Beise, M., Ebling, G., Janz, N., Licht, G. and H. Niggemann (1998): Innovationsverhalten im Verarbeitenden Gewerbe. Ergebnisse der Erhebung 1997. Mannheim.

BMBF (1998): Germany's Technological Performance, 1997 Updated and Expanded Report, Bonn.

Boston Consulting Group (1998): Innovationskraft: Forschende Arzneimittelhersteller am Standort Deutschland.

Büchtemann, Ch.F. and K. Vogler-Ludwig (1997): Das deutsche Ausbildungsmodell unter Anpassungszwang: Thesen zur Humankapitalbildung in Deutschland, in: Ifo-Schnelldienst 17-18/1997, p. 15-20.

Cantwell J. und R. Harding (1998): The Internationalisation of German Companies' R&D, in: National Institute Economic Review, No. 163, p. 99-115.

CPB Netherlands Bureau for Economic Policy Analysis (1997): Challenging Neighbours. Rethinking German and Dutch Economic Institutions, Springer, Berlin, Heidelberg, New York.

DIW (1998): Das Dienstleistungs-Puzzle. Ein aktualisierter deutsch-amerikanischer Vergleich, DIW-Wochenbericht 35/98, p. 625-629.

DIW/IfW/IWH (1998): Gesamtwirtschaftliche und unternehmerische Anpassungsfortschritte in Ostdeutschland. 17. Bericht, Kieler Diskussionsbeiträge 322/323.

Dodgson, M. and R. Rothwell (Ed.) (1994): The Handbook of Industrial Innovation, Edward Elger, Aldershot.

Dosi, G. (1988): Sources, Procedures, and Microeconomic Effects of Innovation, in: Journal of Economic Literature, No. 26, p. 1120-71, London.

Dosi, G., Pavitt, K. and L. Soete (1990): The Economics of Technical Change and International Trade, NYU Press, New York

Ebling, G. and N. Janz (1998): Export Behavior and Innovation Activities in the Service Sector - Empirical Results for a Cross-Section of German Firms, Mannheim.

Ebling, G., Hipp, C., Janz, N., Jungmittag, A., Licht, G. and H. Niggemann (1998): Innovationsaktivitäten im Dienstleistungssektor, Erhebung 1997, Bericht für das BMBF. Mannheim und Karlsruhe.

EU (1995): Green Paper on Innovation, EU-Commission: Luxembourg (http://www.dbs.-cordis.lu)

EU (1997), First Action Plan for Innovation in Europe, EU: Luxembourg (http://-www.dbs.cordis.lu)

FhG-ISI/DIW/ZEW   (1998):   Globalisierung   industrieller   FuE   in   ausgewählten Technikfeldern.

Freeman, Ch. (1994): The Economics of Technical Change, in: Cambridge Journal of Economics, No. 18, p. 463-514.

Freeman, Ch. (1997): The 'National System of Innovation' in Historical Perspective, in: Archibugi, D. and J. Michie (Ed.), Technology, Globalisation and Economic Performance, Cambridge University Press, Cambridge, p. 24-49.

Gehrke, B., Legler, H. and U. Schasse (1998): Regionalökonomische Effekte von Klimaschutzmaßnahmen in der Region Hannover. Studie des NIW im Auftrag des Kommunalverbandes Großraum Hannover.

Gehrke, B., Grupp, H. et al. (1995): Wissensintensive Wirtschaft und ressourcenschonende Technik. Studie des FhG-ISI und des NIW im Auftrage des Bundesministeriums für Bildung und Forschung, Hanover, Karlsruhe, July 1995.

Grömling, M., Lichtblau, K. and A. Weber (1998): Industrie und Dienstleistungen im Zeitalter der Globalisierung, Cologne.

Grossman, G. and E. Helpman (1993): Innovation and Growth in the Global Economy, MIT Press, Cambridge, Mass.

Grupp, H., Hinze, S., Reiß, T. and U. Schmoch (1997): Technologische Position Deutschlands im internationalen Wettbewerb, FhG-ISI, Karlsruhe.

Grupp, H. and H. Legler (1992): Innovationspotential und Hochtechnologie. Technologische Position Deutschlands im internationalen Wettbewerb, Heidelberg.

Härtel, H.H. and R. Jungnickel (1998): Strukturprobleme einer reifen Volkswirtschaft. Analyse des sektoralen Strukturwandels in Deutschland im Auftrage des Bundesministeriums für Wirtschaft, Hamburg, May 1998.

Hermann, C., Konzack, T. and P. Ständert (1998): Analyse zur Entwicklung der Potentiale in Forschung und Entwicklung im Wirtschaftssektor in den neuen Bundesländern im Zeitraum 1990 bis 1997, Studie für den BMWi, Neuenhagen bei Berlin.

International Institute for Management Development (1998): The World Competitiveness Yearbook 1998, Lausanne.

Kern, H. and M. Schumann (1984): Ende der Arbeitsteilung?, Munich.

Kline, S. and Rosenberg, N. (1986): An Overview of Innovation. in: Landau, R. and N. Rosenberg (Ed.), The Positive Sum Strategy: Harnessing Technology for Economic Growth, Washington, D.C..

Klodt, H., Maurer, R. and A. Schimmelpfennig (1997): Tertiarisierung der deutschen Wirtschaft, Institut für Weltwirtschaft, Kiel.

Krugman, P. (1991): Geography and Trade, MIT Press, Cambridge, Mass.

Leadbeater, C. (1999): Living on Thin Air. The New Economy, Viking.

Licht, G. und H. Stahl (1997): Ergebnisse der Innovationserhebung 1996. ZEW-Dokumentation 97-07, Mannheim.

Licht, G., Schnell, W. and H. Stahl (1996): Ergebnisse der Innovationserhebung 1995, ZEW-Dokumentation 96-05, Mannheim.

Milgrom, P. and J. Roberts (1992): Economics, Organisation and Management, Eaglewood Cliffs, N.Y.

Mowery, D. C. and J. Oxley (1997): Inward Technology Transfer and Competitiveness: The Role of National Innovation Systems, in: Archibugi, D. and J. Michie (Ed.), Technology, Globalisation and Economic Performance, Cambridge University Press, Cambridge, p. 138-171.

Nelson, R. R. and S. G. Winter (1982): An Evolutionary Theory of Economic Change, Cambridge.

OECD (1997): OECD Proposed Guidelines For Collecting and Interpreting Technological Innovation Data - OSLO Manual, Second Edition, Paris.

OECD (1998a): Technology, Productivity and Job Creation - Best Policy Practices. Analytical Report, C/MIN(98)7, Paris.

OECD (1998b): Technology, Productivity and Job Creation. Best Policy Practices. Highlights, Paris.

OECD (1998c): Human Capital Investment - An International Comparison, Paris.

OECD (1998d): 1998 Science, Technology and Industry Outlook, OECD: Paris.

Office of Technology Policy (1998): America's New Deficit: The Shortage of Information Technology Workers – Update 1998, U.S. Department of Commerce, Washington, D.C., 1998.

Porter, M. E. (1990): The Competitive Advantage of Nations, Macmillan, New York.

Porter, M.E. and Stern, S. (1999): The New Challenge to America's Prosperity: Findings from the Innovation Index, Council of Competitiveness, March.

Prognos AG (1997/1998): Arbeitslandschaft der Zukunft - Quantitative Projektion für das IAB, Basel.

Prognos AG (1998): Deutschland Report No. 2, Basel.

Soskice, D. (1997): Divergent Production Regimes, in: Kitschelt, H., Lange, P., Marks, G. and Stephens, J.D. (Ed.), Continuity and Change in Contemporary Capitalism, Cambridge University Press.

Stoneman, P. (1995): Handbook of the Economics of Innovation and Technological Change, Edward Elger, Aldershot.

Streeck, W. (1991): On the Institutional Conditions of Diversified Quality Production, in: Matzner, E. and W. Streeck, The Socio-Economics of Production and Employment, London.

Williamson, O.E. (1985): The Economic Institutions of Capitalism, New York.